Analytical Techniques in Animal Nutrition

Analytical Techniques in Animal Nutrition

Authors
S.P. TIWARI, S. RAJAGOPAL
AND
USHA RANI MEHRA

SSPH

SATISH SERIAL PUBLISHING HOUSE
403, Express Tower, Commercial Complex, Azadpur, Delhi-110033 (INDIA)
Phone : 011-27672852, Fax : 91-11-27672046
E-mail : info@satishserial.com, hkjain1975@yahoo.com
Website : www.satishserial.com

Published by :

SATISH SERIAL PUBLISHING HOUSE
403, Express Tower, Commercial Complex, Azadpur, Delhi-110033 (INDIA)
Phone : 011-27672852 Fax : 91-11-27672046
E-mail : info@satishserial.com, hkjain1975@yahoo.com

Reprint - 2014

ISBN 978-9381226-01-8

Typeset at: Harminder *for* Laxmi Art Creation
Printed at: Salasar Imaging Systems, Delhi-110035

PREFACE

There are some laboratory manuals on Animal Nutrition and feeding written by eminent teachers and scientists. But no single manual on Animal Nutrition caters the needs of undergraduate and post graduate students of Veterinary science by providing all informations at a single source. This book comprises of principles of animal nutrition, evaluation of feed stuffs and feed technology and applied animal nutrition covering feeding of livestock, poultry, human beings, pet, rabbit and laboratory animals. This book contains the informations as required in the new syllabus and examination system approved by the Veterinary Council of India.

Book consists of two Sections (Section-I and Section-II). Both Sections contain two parts (Part-I and Part-II) and thus four parts correspond to four courses. Each Sections provides a structured approach to learning by covering 'all the practical topics in a uniform, systematic format. This book is designed to give the students rapid, easy access to all the material in a course wise format which benefit them prepare for the internal examinations and external examinations economically and ensure exploit their potential to a greater extent. The topics are detailed in a straightforward and hopefully lucid manner.

Each exercise is introduced with the text material which will orient the students with the subject. Brief stepwise instructions have been given to perform the exercises. Wherever required, tables, blank spaces or charts have been provided to note down the results and observations. Although the calculations were made more easy by putting the hypothetical observations. The exercise after the chapters will stimulate the students to seek more informations on the subject. Formulae and details of the preparation of certain reagents, *etc.*, are given in the text.

By using this book as a guide, if the students get a clear understanding and a systematic and methodical approach to the subject, the purpose of writing this manual is accomplished. The authors earnestly hope that the book will meet the long felt demand of such book in this country. The authors are thankful to Dr. Kuldeep Deshpandey and Dr. Kiran Kumari, Ph.D. scholars who have helped in the editing of this manual.

Authors

CONTENTS

SECTION- II
Part - I

SECTION-I
Part-I

PRINCIPLES OF ANIMAL NUTRITION INCLUDING AVIAN

1. General Precautions While Working in Animal Nutrition Laboratories

General Instructions

Some general instructions for working in the laboratory are given here and the students are advised to read them carefully before they begin their work in the laboratory.

1. Apparatus used should be kept neat and clean.
2. While pipetting rubber bulb should be used.
3. Small amount of liquid remaining in the pipette should not be blown out. In standardizing the pipettes, allowance is always made for this. However, a frosted ring near the top of latest designed pipettes indicates that the last drop is to be blown out.
4. Most measuring devices are made of glass which has a small temperature coefficient, eg: a soft glass vessel will lead to a change in volume by about 0.003 per cent per degree (°C) change in

temperature; with heat resistant glass the change is about one-third of this. As a general rule, heating of calibrated equipment should be avoided. Rapid cooling can change the glass structure and causes change in volume.

Laboratory Safety

1. Laboratory should be well ventilated and fitted with exhaust fans for effective removal of fumes.
2. Use apron and other devices like gloves, goggles etc. depending upon the material to be handled.
3. Do not add water to acids. Keep acids off skin and protect eyes from spattering. If acids are spilled on the skin, wash off immediately with tap water. Gaseous nitrogen oxides from HNO_3 can cause severe lung damage. Copious flow of fumes occurs when both concentrated HNO_3 and HCl are mixed together.
4. If acid falls on clothes, neutralize the same with few drops of dilute ammonia solution or some other weak alkali solution.
5. If acid spills on the floor or the table, neutralize it with some weak alkali and wipe off with duster.
6. If you happen to suck acid into your mouth during pipetting wash your mouth quickly with water and then rinse with a weak solution of washing soda.
7. Use fume hood to protect against any type of fumes.
8. Avoid use of equipment for purposes other than intended.

Cleaning Laboratory Ware

The glass and porcelain ware should be thoroughly washed with some detergent and should be extensively rinsed with tap water followed by further rinsing with distilled water.

If a grease film remains after cleaning with detergent, a cleaning solution consisting of sodium or potassium dichromate in concentrated sulphuric acid may be used. After this, rinsing is necessary in order to remove the last traces of dichromate ions which adhere strongly to glass or porcelain surface.

Preparation of Cleaning Solution

Weigh 60g of potassium dichromate and transfer to a 1000 ml pyrex beaker, add 300ml tap water, and mix thoroughly with glass rod. Heat the solution to boil till potassium dichromate dissolves. Allow the solution to cool and then add slowly 640ml of commercial grade sulphuric acid while mixing thoroughly with glass rod. Transfer the contents to 1 litre capacity cylinder for cleaning glassware.

> **Note**
>
> 1. Cleaning solution should be discarded when it acquires green colour of chromium ion.
> 2. Cleaning solution is most effective when warmed up to about 70°C. At this temperature it rapidly attacks plant and animals matter and thus it is potentially dangerous preparation.
> 3. Spillage, if occurs, should be cleaned with water.

Commonly Used Glassware

Beaker Volumetric flask Conical flask Erlenmayer

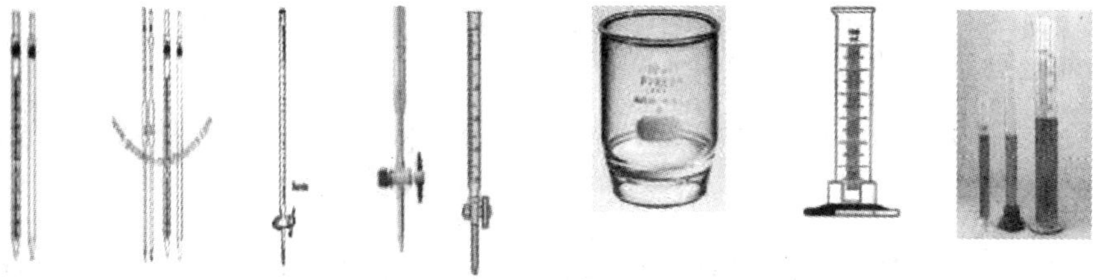

Pipettes Burettes Sintered crucible Measuring cylinder

Exercise

1. Familiarise yourself with the commonly used glassware in Animal Nutrition laboratory.
2. Prepare 100ml of cleaning solution.

2. Standard Solutions

Standard solution : A standard solution is the solution of known strength.

Normal solution : A normal solution contains one gram-equivalent weight of the solute per litre of the solution. Normality of a solution is usually designated as N.

Determination of Gram-equivalent Weight of an Acid and an Alkali

Equivalent weight of a substance is its weight equivalent in reacting power to an atom of hydrogen. It is calculated as follows:

$$\text{Equivalent weight of an acid} = \frac{\text{Molecular weight of an acid}}{\text{Basicity}}$$

Where, basicity of an acid is equal to the number of replaceable hydrogen atoms present in one molecule of the acid. Thus, equivalent weight of H_2SO_4 is 98/2=49 and of HCl is 36.46/1 = 36.46.

$$\text{Equivalent weight of an alkali} = \frac{\text{Molecular weight of an alkali}}{\text{Acidity}}$$

Where, acidity of an alkali is equal to the number of replaceable hydroxyl groups present in one molecule of alkali. Thus, equivalent weight of NaOH is 40/1= 40.

Equivalent weight expressed in gram is called gram-equivalent weight.

Determination of Gram-equivalent Weight of Oxidizing Agents

Gram-equivalent weight of an oxidizing agent is the weight of the substance which is equivalent to 8 g of available oxygen.

Equivalent weight of $KMnO_4$ in acidic medium is determined as follows:

$$2\ KMnO_4 + 3\ H_2SO_4 = K_2SO_4 + 2MnSO_4 + 3H_2O + 5(O)$$

$$2(39 + 55 + 64)$$

$$(2 \times 158) \text{ or } 316 = 5(O) = (5 \times 16) \text{ or } 80$$

Hence, 316 g of $KMnO_4$ will yield 80g of oxygen for oxidation. Therefore, 8g of oxygen will be displaced by 31.6 g of $KMnO_4$.

Equivalent weight of $KMnO_4$ in alkaline medium is determined as follows:

$2\ KMnO_4 + H_2 = 2MnO_2 + 2KOH + 3(O)$

In this case, 316 g of $KMnO_4$ will yield 48g of oxygen for oxidation

$2KMnO_4 = 3(O) = 6H$

So, equivalent weight of $KMnO_4$ in alkaline medium is :

$$\frac{316}{48} \times 8 = 52.67$$

Molar Solution

Molar solution is one that contains one mole or one molecular weight in grams of a substance in each litre of the solution, whether the substance is in the form of molecules, ions or any other species. It is designated as M.

Molar method of expressing concentration is useful due to the fact that equal volumes of equimolar solutions contain equal number of molecules.

Molarity

Number of moles of a substance per liter of solution is called molarity.

Molality

Molality of solution is the number of moles of solute per 100g of the solvent. It is designated as 'm'.

Titration

A form of analysis by means of comparision with solutions of standard strength.

Acid-Base Indicators

Acid-base indicators are those substances the colour of which in a solution is dependent on the pH of the medium. They are generally complex organic compounds of fairly high molecular weight. In water or other solvents they behave as weak acids or bases and thus participate in equilibrium reactions involve hydrogen ions. The change of colour is due to internal structural re-arrangement which is responsible for indicator property.

Indicator : H $\rightarrow$ H^+ + Indicator $^-$
(Acid)
Indicator : OH $\rightarrow$ OH^- + Indicator $^+$
(Base)

Some of the important titration and choice of indicators are given here under :

pH range	Indicator	Acid colour	Base colour
3.1 to 4.4	Methyl orange	Pink	Yellow
4.2 to 6.2	Methyl red	Red	Yellow
6.0 to 7.6	Bromothymol blue	Yellow	Blue
8.2 to 10.0	Phenolphthalein	Colourless	Pink
10.0 to 12.0	Alizarine yellow	Yellow	Violet

Use of Indicators

Acid	Alkali	Indicator to be used
Strong	Strong	Any indicator can be used
Strong	Weak	Methyl orange (or) Methyl red
Weak	Strong	Phenolphthalein

Titre Endpoint

It is the point at which the reaction is complete i.e. at which complete interaction between an acid and an alkali in equivalent amount takes place. This end point can be found by a change in colouration of the indicator in the solution.

Standard Solutions

There are a few standard solution which are used in proximate analysis of feed stuffs. These are :

1. N/10 H_2SO_4
2. N/10 NaOH
3. N/10 $KMnO_4$
4. 1.25% H_2SO_4 (w/v)
5. 1.25% NaOH (w/v)
6. 40% NaOH (w/v)

Certain primary standard solutions are also required for standardization of the above solutions. These are :

1. N/10 Na_2CO_3
2. N/10 $(COOH)_2.2H_2O$

3. Preparation of Standard Solution

In this chapter procedures for the preparation of commonly used standard solutions in Animal Nutrition laboratory are furnished.

Preparation of N/10 Na_2CO_3 Solution

Calculations

Molecular weight of Na_2CO_3 = 106

$Na_2CO_3 + 2HCl = 2NaCl + H_2O + CO_2$

So, acidity of Na_2CO_3 = 2

Equivalent weight of Na_2CO_3 = 106/2 = 53

Therefore, 5.3 g Na_2CO_3 is required for each litre of solution to make N/10 Na_2CO_3. Since it is hygroscopic, it must be made perfectly anhydrous before it is weighed.

Procedure

Take 6-7 g of Na_2CO_3 (AR grade) in a crucible and heat it in a hot air oven at about 100°C for few hours so as to drive out any moisture and to convert any preformed $NaHCO_3$ to Na_2CO_3. Cool in a desicator and weigh exactly 5.3g dried salt and dissolve it in a little quantity of freshly boiled distilled water. Transfer it to one litre measuring flask and make volume up to the mark. Shake well and label it as 0.1N Na_2CO_3 solution.

Preparation of N/10 Oxalic Acid

Calculations

Oxalic acid [$(COOH)_2$. $2H_2O$] has molecular weight of 126 and basicity 2. So, equivalent weight of oxalic acid = 126/2 = 63

Thus, 6.3g of $(COOH)_2$. $2H_2O$ is to be dissolved in one litre of distilled water to get N/10 oxalic acid solution.

Procedure

Weigh accurately 6.3g $(COOH)_2 . 2H_2O$ and transfer it to a volumetric flask (1 litre) half filled with distilled water. Shake well and make the volume upto the mark. Label it as N/10 oxalic acid solution.

> **Note**
>
> If anhydrous oxalic acid $(COOH)_2$ is available then dissolve 4.5g of the acid in one litre of distilled water to get 0.1 N oxalic acid solution.

Preparation of N/10 H_2SO_4 Solution

Calculations

Equivalent weight of H_2SO_4	=	49
Specific gravity	=	1.84
So, volume of 49g H_2SO_4	=	26.6 ml (*i.e.* 49/1.84)

Concentrated H_2SO_4 (reagent grade) is about 97% pure.

Therefore, actual amount of concentrated H_2SO_4 required for one litre of 1N H_2SO_4 solution = 100/97 x 26.6 = 27.42 ml. Thus, for one litre of N/10 H_2SO_4 solution, 2.74 ml of concentrated H_2SO_4 is required.

Procedure

Take 2.74 ml sulphuric acid in a beaker half filled with distilled water. Transfer the contents and washing to a volumetric flask (1 litre) and make volume upto the mark. Shake well and titrate this solution with 10ml of 0.1 N Na_2CO_3 using methyl orange as indicator. Repeat the titration to get at least three concordant readings.

Standardisation

Suppose 10 ml of 0.1N Na_2CO_3 = 9.5 ml of N/10 H_2SO_4

$V_1N_1 = V_2N_2$ $\qquad$ $N_1V_1 = N_2V_2$

Or, $10 \times 0.1\ N = 9.5 \times N_2$ $\qquad$ $0.1 \times 1000 = 0.10526 \times V_2$

Or, $N_2 = 0.10526$

To prepare 1 litre N/10 H_2SO_4, the volume of 0.10526 N acid required is 1000 x 0.1/0.10526 = 950 ml of 0.10526 N acid which should be diluted to one litre. Check it again with N/10 Na_2CO_3 for three times. It must neutralize equal volume of N/10 Na_2CO_3 solution. Label it as 0.1N H_2SO_4 solution.

Precaution : Never add water to an acid.

Molecular weight of NaOH	=	40
Acidity (number of replaceable OH group)	=	1
Equivalent weight of NaOH	=	40

Therefore, 4g of NaOH dissolved in one litre of solution will give N/10 NaOH solution.

Procedure

Weigh quickly 4g of NaOH in a beaker (as it is hygroscopic) and dissolve it in distilled water. Transfer the contents and the washings to a volumetric flask (1 litre). Cool and then make volume upto the mark. Shake well and standardize this solution against N/10 oxalic acid using phenolphthalein as an indicator. Label it as 0.1 N NaOH solution.

Preparation of N/10 $KMnO_4$ Solution

Dissolve 3.2 g of $KMnO_4$ in one litre of distilled water. Boil it for 10-15 minutes and then allow to stand for few days and then filter it through glass wool.

Take 10ml of N/10 oxalic acid in a beaker. Add 5 ml of dilute sulphuric acid warm it to 60-70°C and titrate against $KMnO_4$ from the burette till a light pinkish colour appears. Take three concordant readings.

10 ml 0.1 N oxalic acid = 9.75 ml of $KMnO_4$

$$10 \times 0.1\ N = 9.75 \times N_2$$

$$N_2 = \frac{10 \times 0.1\ N}{9.75}$$

$$= 0.10256$$

To prepare 1000 ml 0.1N $KMnO_4$ solution the volume of $KMnO_4$ to be taken:

$$\frac{1000 \times 9.75 \times 0.1}{10 \times 0.1} = 975\ ml$$

Now take 975 ml of prepared $KMnO_4$ solution and make it to 1000 ml by adding distilled water.

Note

1. Ordinary or even pure distilled water contains traces of organic matter which reduces the strength of $KMnO_4$ solution. That is why the solution is boiled and kept for some time before standardization.
2. In the absence of sufficient amount of dilute H_2SO_4 or due to the rapid addition of $KMnO_4$ in titration flask, brown turbidity (manganous oxide) may appear.

Preservation of Standard Solutions

The bottle must be kept tightly stoppered to prevent evaporation of solvent.

Some solutions must be protected from atmospheric gases. For example, sodium hydroxide solution is affected by atmospheric CO_2.

$$2NaOH + CO_2 = Na_2CO_3 + H_2O$$

KMnO4 solution should be preserved in coloured (amber) bottles.

The container should be shaken well before the withdrawl of a portion of solution to ensure uniform composition of the solution.

Exercise

1. Prepare 100 ml of N/10 Sodium carbonate standard solution
2. Prepare 100ml of N/10 oxalic acid standard solution
3. Prepare 1 litre of N/10 sulphuric acid standard solution.
4. Prepare 1 litre of N/10 sodium hydroxide standard solution.
5. Prepare 1 litre of N/10 potassium permanganate standard solution.

4. Preparation of Common Reagents and Indicators

The following two reagents are used in proximate analysis in addition to the previously mentioned standard solutions.

A. Preparation of Reagents

1. Preparation of 1.25% (w/v) H_2SO_4 solution

Calculations

To prepare 1.25% (w/v) H_2SO_4 solution, 12.5 g of H_2SO_4 (100%) is to be added to distilled water to make up the volume to 1000ml.

$$\text{Volume of } H_2SO_4 \text{ to be taken} = \frac{12.5 \times 100}{1.84 \times 97} = 7 \text{ ml}$$

Procedure

Add 7.0 ml of concentrated H_2SO_4 (specific gravity 1.84 and 97% concentration) in a 1000ml volumetric flask half-filled with distilled water. Shake well and add distilled water to make volume upto the mark.

Preparation of 1.25% NaOH solution

Add 13.16 g of NaOH (95% NaOH) in one litre distilled water and shake well.

B. Preparation of Indicators

1. *Methyl red :* It is saturated solution in 50% absolute alcohol.
2. *Methyl orange :* 0.5% solution in distilled water.
3. *Methylene blue :* Dissolve one standard tablet of Methylene blue (BDH) in 200ml of previously boiled water. Keep in a coloured bottle in a cool and dark place.
4. *Methyl violet or Gentian violet :* It is a 1% solution in 50% absolute alcohol.
5. *Phenolphthalein :* It is a 1% solution in absolute alcohol.
6. *Bromo-cresol (green or purple):* It is a 0.1% solution in alcohol.

Exercise

1. Prepare 1 litre of 1.25% solution of H_2SO_4
2. Prepare 1 litre of 1.25% solution of NaOH

5. Preparation of Samples for Chemical Analysis

The procedure used in the preparation of samples depends upon the purpose of analysis and the nature of the constituents that are to be determined.

Sampling of Wet Material

a) Plant Material (Herbage)

1. Separate the plant material by using a sharp knife.
2. Remove the contaminations like soil or other material by gentle brushing. Never wash the plant material with water.
3. Mix corresponding plant parts from more than one plant to obtain representative sample of sufficient size for analysis.
4. Cut the sample immediately into small pieces and mix thoroughly.
5. Weigh a portion of this sample for those analysis which are to be performed on green material.
6. Dry the sample in a hot air oven at 100 ± 0.5°C for storage.

b) Silage

1. Collect silage material from different parts of silo pit in a polythene bag.
2. Weigh fresh sample for the analysis of nitrogen and dry matter.
3. Prepare silage extract for the determination of pH, volatile fatty acids, (VFAs) and ammonia nitrogen (NH_3-N).
4. Preserve dried ground samples for determination of different proximate principles.

c) Cattle Faeces

1. Mix daily collection thoroughly on a clean surface by hand.
2. Sub-sample 1% of wet weight for dry matter determination.
3. Sub-sample 0.5% of wet weight and preserve in bottle with 40% (w/v) H_2SO_4 after thorough mixing for nitrogen determination.
4. Store the dried and ground faeces in bottles for further analysis.

d) Sheep / Goat Faeces

1. Mix the faeces pellets thoroughly and sub-sample 5% and 1% for dry matter and nitrogen analysis, respectively as described earlier.

e) Poultry Excreta

1. Collect excreta on polythene sheet.
2. Separate feathers, feed and other contaminations.
3. Mix thoroughly and sub-sample 5% and 1% separately for dry matter and nitrogen determinations. For nitrogen analysis, samples should be preserved with 5% (w/v) sulphuric acid.
4. Preserve dried and ground samples for further analysis.

Sampling of Dry Material

a) Hay or Straw

1. Chop hay or straw to pieces of about 3.5 cm.
2. Dry hay or straw at a temperature below 65°C.
3. Grind it through 2mm screen in a mill and preserve in plastic bags for analysis.

b) Concentrates

1. Take representative samples from different parts of each bag.
2. Keep for dry matter estimation in hot air oven at 100 ± 0.5°C.
3. Grind and preserve for further analysis.
4. Label the containers clearly and correctly.

Sampling for Chemical Analysis

1. Draw sample from different places to obtain a representative sample for analytical work.
2. Weigh sample simultaneously for moisture determination to avoid error caused due to atmospheric humidity.

Sampling

The following are the approximate quantities of samples of feed/fodder to be taken for proximate analysis.

a) Moisture Determination

1. 10g of concentrates and dry roughages
2. 50 to 100g of green fodders and leaves

b) Total Ash Determination: Same as Above (a)

c) Ether Extract Determination

1. 2g of oil seeds and oil cakes.
2. 3 to 5g of cereal grains and pulses.
3. 6 to 10g of dry roughages like straw and hay.

d) Crude Fibre Determination

1. 2g of fat and moisture free sample.

e) Crude Protein Determination

1. 0.5 to 1g of oil cakes/protein supplements.
2. 1.5 to 2g of grains.
3. 2 to 3g of dry fodders.
4. 10g of green fodders and leaves.

EXERCISE

1. Practice the sampling and weighing of various feeds and fodders that are commonly used for animal feeding.

LAB EQUIPMENTS AND INSTRUMENTS

Soxhlets apparatus

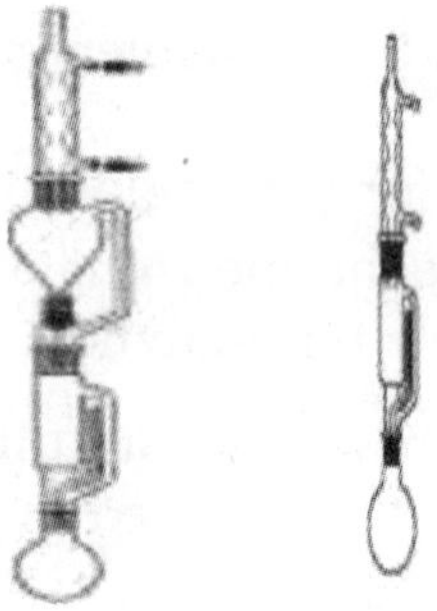

Extractor (condenser-extractor-oil flask)

Dessicator

N-distillation microkjeldahl assembly

Water distillation Assembly

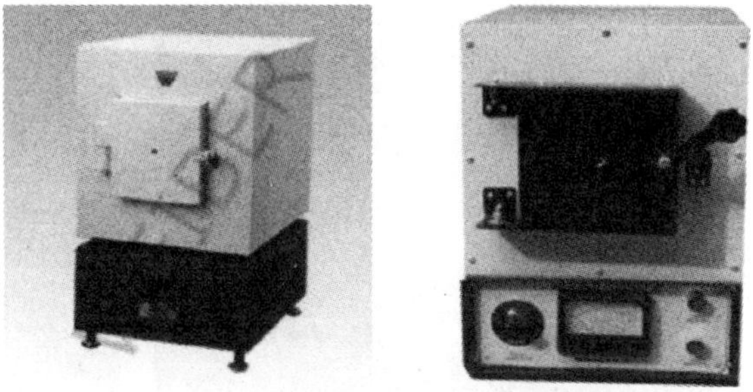

Muffle furnace

Single pan electronic balance

Chemical balance

Physical balance

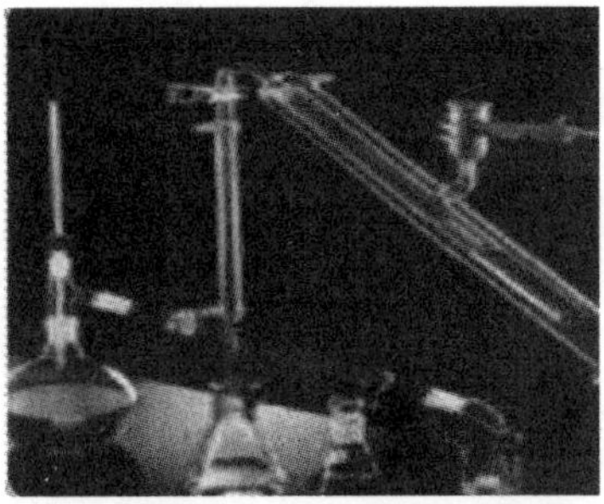

N-distillation macrokjeldahl assembly

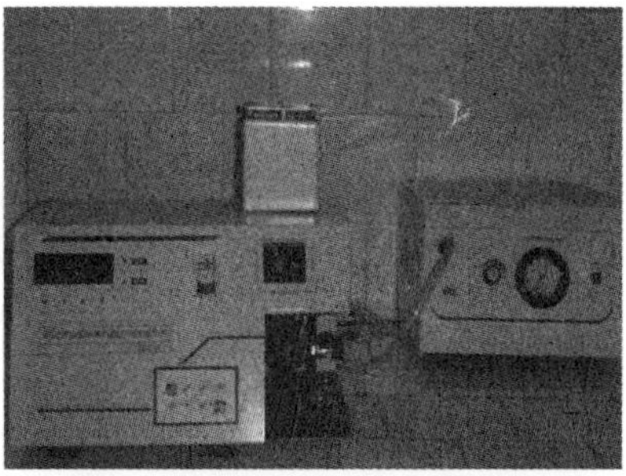

Flame photometer

Bomb Calorimeter

Atomic absorption spectrophotometer

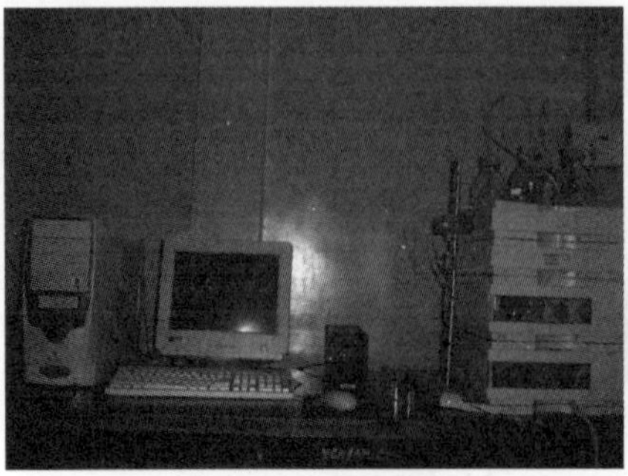

HPLC (High performance liquid chromatography)

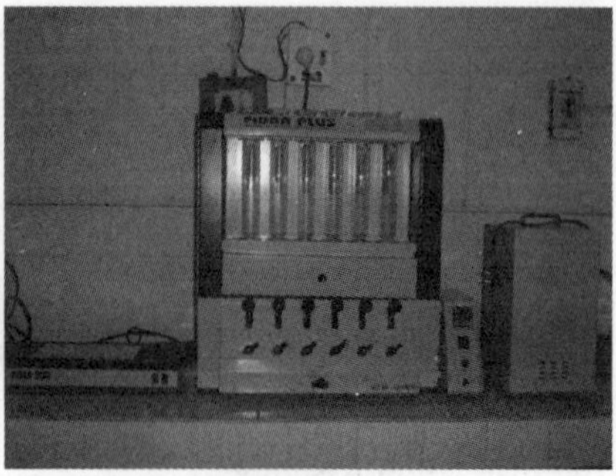

Fibra plus

Socs plus

Willy mill

Hot air oven

Water bath

Hot plate

Heating mantle

Lab autoclave (horizontal)

Ball mill

Hot plate

Vacuum pump

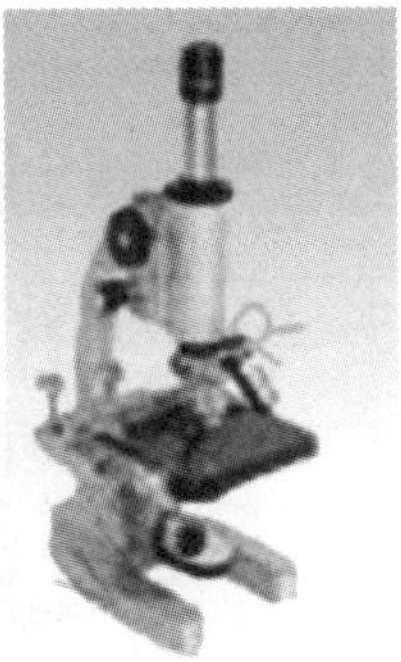

Monocular microscope

pH meter (digital)

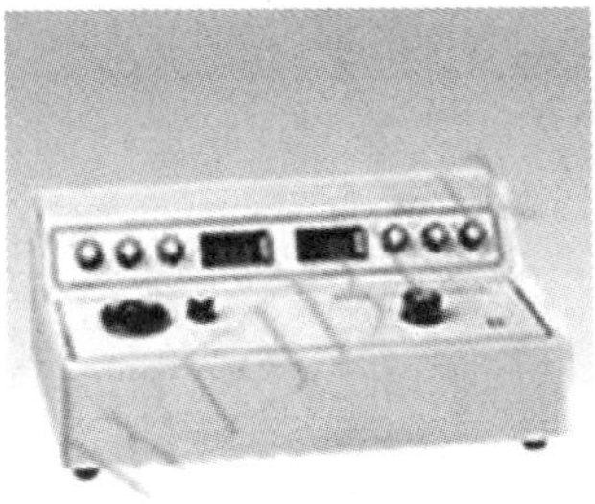

Spectrophotometer

Freeze drier (Lyophiliser)

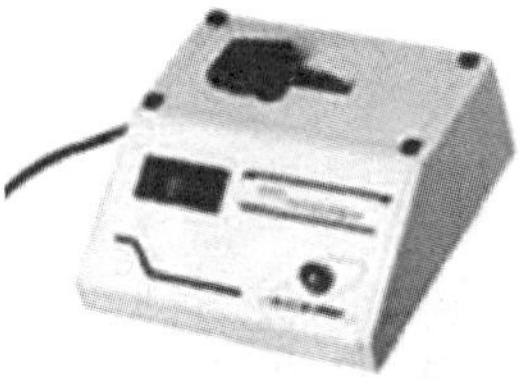

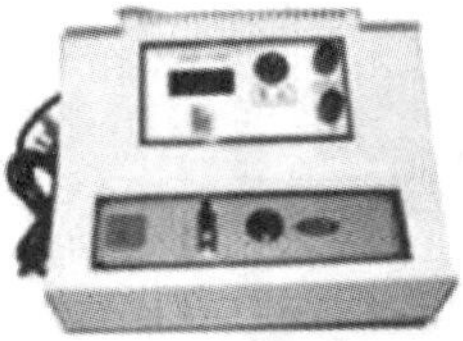

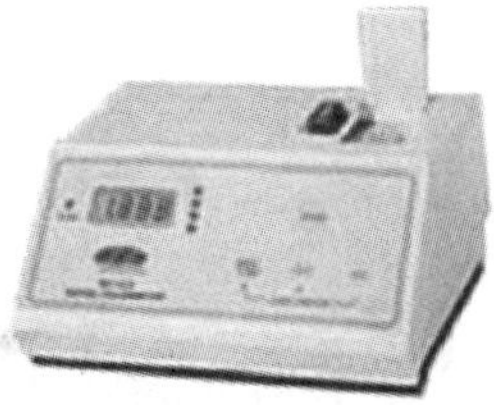

Photocolorimeter

High speed centrifuge

Refrigerated centrifuge

Infra red moisture balance

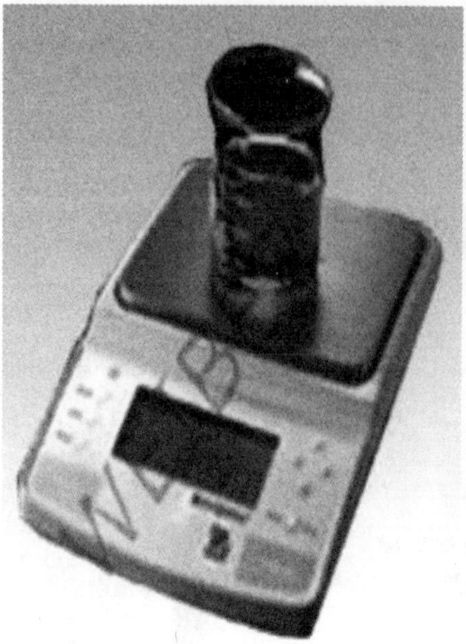

Electronic top loading balance

Electronic plateform balance

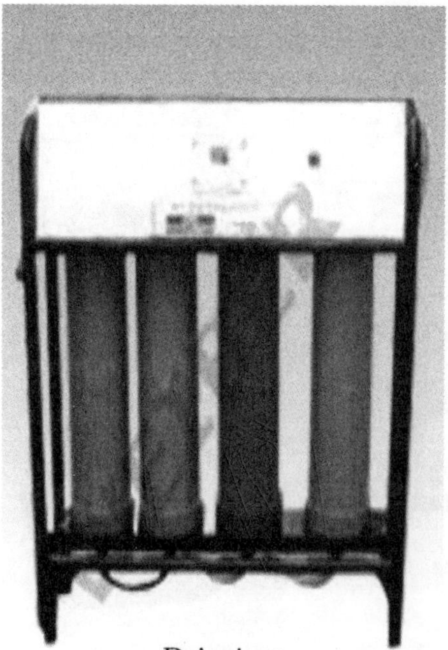

Deioniser

BOD incubator

6. Proximate Analysis of Feeds and Fodders

Proximate system of analysis was developed by Henneberg and Stohmann in the year 1860 in Weende Experimental Station near the University of Goettingen in Germany. Therefore, this system is also called as Weende system of analysis.

The Features of Weende System of Analysis

It is a system for approximating the nutritive composition of a feed. The principle of the analysis is to separate feedstuffs into various fractions. Eg.: water or moisture, crude protein, ether extract, crude fibre, nitrogen-free extract and mineral matter or total ash. The total mineral matter represents the inorganic part of the material, whereas crude protein, ether extract, nitrogen-free extract and crude fibre are collectively called organic matter.

7. Determination of Dry Matter

The constant weight of a sample after complete removal of moisture or water is generally called dry matter (DM% = 100 - moisture).

Different methods of dry matter (moisture) determination are :

1. Oven drying
2. Toluene distillation
3. Freeze drying
4. Infra-red moisture meter.

Oven Drying Method

Principle

The moisture content of a sample is determined by heating it in an oven to a constant weight at 100-105°C under atmospheric pressure. The moisture is removed as "vapours".

Apparatus

Balance moisture cup with lid, hot air oven, desiccators, asbestos sheet and metal tongs.

Procedure

1. Dry the moisture cup in an oven at 100°C, cool it in a desiccator and record its weight.
2. Take about 10g of the material in the moisture cup and weigh out to calculate actual amount of material taken.
3. Dry it in a hot air oven with lid partly open at 100°C for 8 hours (less than 100°C with forced draft oven or under vacuum till a constant weight is achieved).
4. Remove the moisture cup from the oven and cool it in a desiccator.
5. Repeat the process of heating and cooling till a constant weight is achieved and record the constant weight of moisture cup with sample.

Calculations

$$\text{Moisture (\%)} = \frac{\text{Initial wt of moisture cup- Final wt after drying + sample}}{\text{Wt. of sample before drying}} \times 100$$

$$\text{Dry matter (\%)} = \frac{\text{Wt. of dried sample}}{\text{Wt. of sample before drying}} \times 100$$

Or, Dry matter = 100 - % moisture

Precaution

Weighing should be quick and accurate.

Exercise

1. Determine the dry matter content of the given feed sample and enter the readings and your observations in the record book.
2. Calculate the following:
   a. Weight of empty cup is 26.8125 g. Weight of the cup with sample is 35.2753 g. Weight of the cup after evaporation of moisture is 28.6543. Calculate the dry matter content of the sample.
   b. If the calcium content in a 1.0g feed sample is 0.2 g on 75% moisture basis. What would be the calcium content on dry matter basis.

c. The protein and dry matter content of maize from source A is 8.5 and 85 and that of maize from source B is 8 and 92%. Which maize sample is richer in protein content on dry matter basis?

8. Determination of Total Ash Content

Principle

Ash is the inorganic or mineral component of the sample left after its complete ignition at 600°C in a muffle furnace.

Apparatus

Oven, balance, silica crucible, metal tongs (long), muffle furnace, asbestos sheet and desiccator.

Procedure

1. Dry the crucible in an oven, cool it in a desiccator and record its weight.
2. Weigh about 5g of oven dried sample into a crucible.
3. Ignite the sample on a burner or on an electric heater initially till the smoke ends up.
4. Transfer the crucible to a muffle furnace with the help of a pair of long tongs. Keep it at 600°C for two hours. (At this temperature all the organic matter will be burnt leaving white ash).
5. Remove the crucible from the furnace, keep it on an asbestos sheet, cool it in a desiccator and weigh. The previously recorded empty crucible weight is now subtracted and the weight of ash is thus determined.

Preserve this ash for extraction of the acid soluble minerals and, thus for estimation of the acid insoluble ash.

Calculation

$$\text{Ash or mineral matter (\%)} = \frac{\text{Wt. of ash (g)}}{\text{Original wt. of sample (g)}} \times 100$$

Precaution

The ash is highly hygroscopic and thus weighing should be done quickly.

Exercise

1. Determine the total ash content of given feed sample and record your observations in your practical record.
2. Weight of the sample taken is 8.2450 g. Weight of the silica crucible after ashing is 36.5437g. Weight of the empty crucible is 35.1235g. Calculate the total ash content of the sample.
3. Weight of the sample taken is 10.6540 g. Weight of the silica crucible after ashing is 40.0150 g. Weight of the empty crucible is 38.8245g. Calculate the organic matter content of the feed sample.

9. Determination of Acid Insoluble Ash

Principle

The ash is digested with dilute HCl (1:1) to dissolve inorganic salts.

Apparatus

Whatman filter paper No. 42 or its equivalent (No. 40, 41, 44 *etc.*), glass funnel, volumetric flask, watch glass, pipette, desiccator, balance and muffle furnace.

Reagents

Concentrated HCl and 1 or 2 drops of HNO_3.

Procedure

1. Moisten the ash sample which was prepared in the previous experiment by using little distilled water.
2. Add 10ml of concentrated HCl in the crucible containing ash.
3. Add 1 or 2 drops of concentrated HNO_3 and mix thoroughly with glass rod.
4. Place the crucible on an electric heater and boil the contents for a while.
5. Cool the contents while mixing the contents thoroughly.
6. Filter through Whatman filter paper (No. 42) into a 100ml volumetric flask, wash and transfer the residue in the crucible repeatedly with small volumes of hot distilled water.

7. Make up the volume to the mark of the volumetric flask. Mix the contents thoroughly and preserve this solution for the estimation of different inorganic ingredients.
8. Transfer the filter paper along with the residue to the same crucible and ignite it in a muffle furnace at 600°C.
9. Cool the crucible in a desiccator and record its weight. The increase in weight of crucible gives the weight of acid insoluble ash.

Calculation

$$\text{Acid insoluble ash (\%)} = \frac{\text{Wt. of insoluble ash (g)}}{\text{Original wt. of sample (g)}} \times 100$$

Precaution

Do not make volume in the flask while its contents are hot.

Exercise

1. Determine the acid insoluble ash content of the feed sample and give your inference with regard to the quality of the feed ingredient.
2. Calculate the acid insoluble ash content of the sample mentioned in question no. 2 and 3 of the previous exercise by using the following data:

   a. Weight of silica crucible with acid insoluble ash is 35.4350 g.

   b. Weight of silica crucible with acid insoluble ash is 39.1045 g.

10. Familiarisation of Various Feed Stuffs

Identify the feed and fodders that are commonly used for feeding of livestock. Write the significance of each in animal feeding point of view alongwith their nutritive values.

Feed Ingredients

S.No.	Name	Feed ingredients	Nutritive value	
1.	Soybean	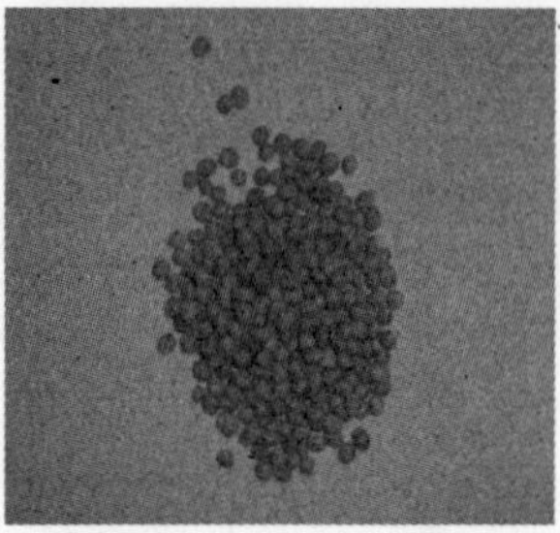	ME (Kcal/kg)	2400
			Protein	45%
			Crude fibre	6%
			Ca	0.3%
			Available P	0.28%
			Lysine	2.79%
			Methionine	0.65%
			Meth + Cystine	1.32%
			Linoleic acid	0.52%
2.	Yellow maize		ME (Kcal/kg)	3350
			Protein	9.0%
			Crude fibre	3.0%
			Ca	0.02%
			Available P	0.09%
			Lysine	0.26%
			Methionine	0.19%
			Meth + Cystine	0.39%
			Linoleic acid	1.50%
3.	Deoiled rice bran	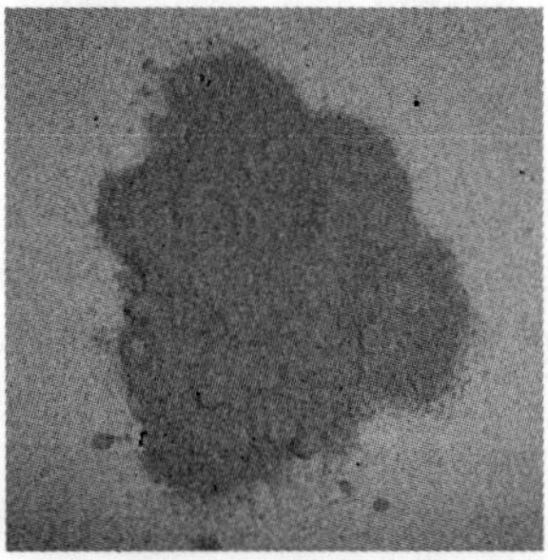	ME ME (Kcal/kg)	2000
			Protein	14.0%
			Crude fibre	16.0%
			Ca	0.07%
			Available P	0.21%
			Lysine	0.64%
			Methionine	0.29%
			Meth + Cystine	0.59%
			Linoleic acid	0.26%
4.	Maize gluten		ME ME (Kcal/kg)	2500
			Protein	64.0%
			Crude fibre	6.0%
			Ca	0.02%
			Available P	0.09%
			Lysine	1.89%
			Methionine	1.36%
			Meth + Cystine	2.80%
			Linoleic acid	2.20%

5. Ground nut cake

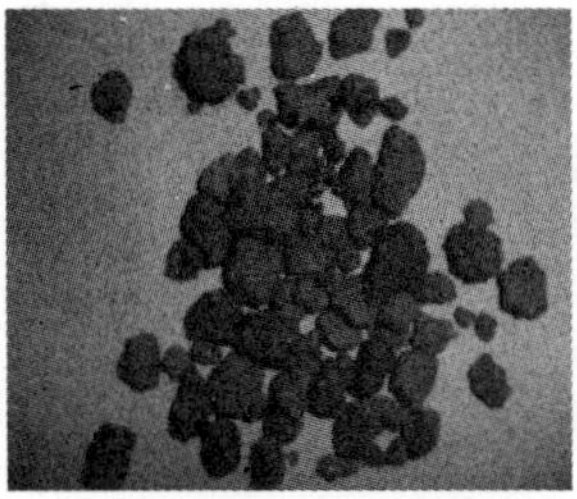

ME ME (Kcal/kg)	2300
Protein	45.0%
Crude fibre	7.0%
Ca	0.30%
Available P	0.15%
Lysine	1.64%
Methionine	0.52%
Meth + Cystine	1.16%
Linoleic acid	0.33%

6. Rape seed meal

ME ME (Kcal/kg)	2100
Protein	38.0%
Crude fibre	12.0%
Ca	0.60%
Available P	0.30%
Lysine	1.92%
Methionine	0.72%
Meth + Cystine	1.75%
Linoleic acid	0.16%

7. Fish meal

ME ME (Kcal/kg)	2500
Protein	45.0%
Ca	8.0 %
Available P	2.0%
Lysine	3.59%
Methionine	1.33%
Meth + Cystine	1.69%
Linoleic acid	0.09%

8. Blood meal

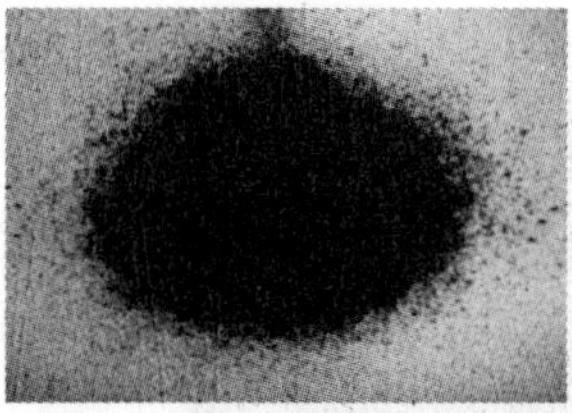

ME ME (Kcal/kg)	3000
Protein	86.0%
Crude fibre	0.70%
Ca	0.32%
Available P	0.31%
Lysine	7.0%
Methionine	0.83%

9. Meat and bone meal

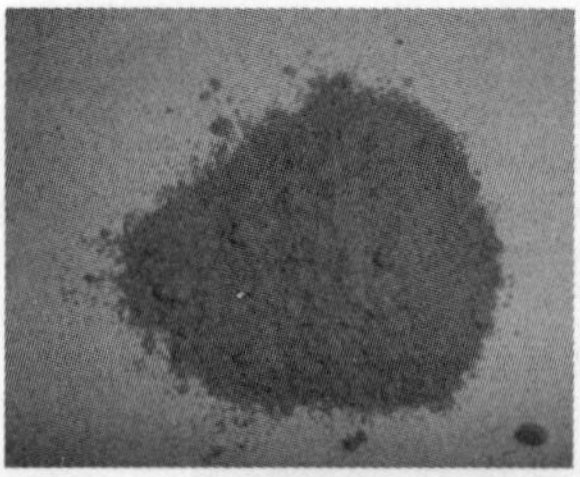

ME ME (Kcal/kg)	2700
Protein	48.0%
Ca	10.0%
Available P	4.0%
Lysine	2.43%
Methionine	0.65%
Meth + Cystine	1.16%
Linoleic acid	0.40%

10. Rice polish

ME ME (Kcal/kg)	3000
Protein	16.0%
Crude fibre	11.2%
Ca	0.37%
Available P	1.80%

11. Dicalcium phosphate

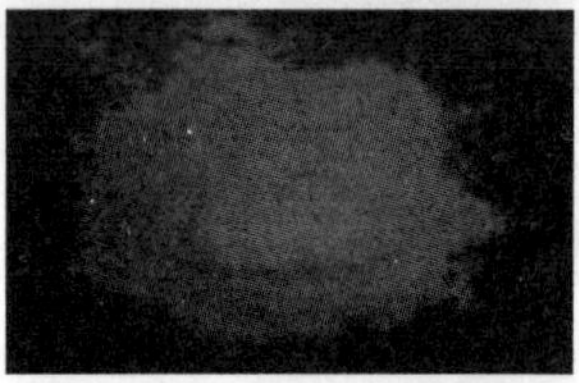

Ca	24 %
Total P	17 %
Available P	16 %

12. Lime stone powder

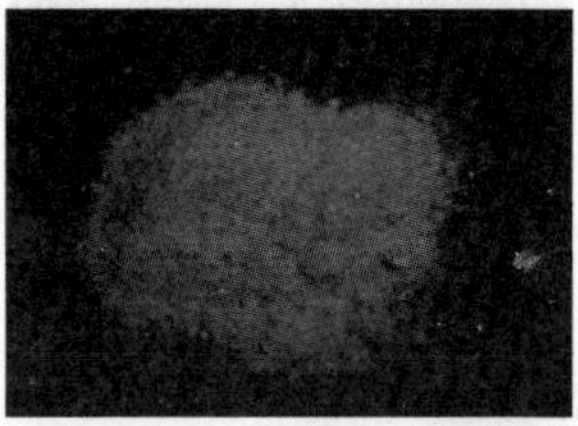

Ca	38 %

13. Wheat straw

TDN	40 %
CP	3%

14. Soybean husk

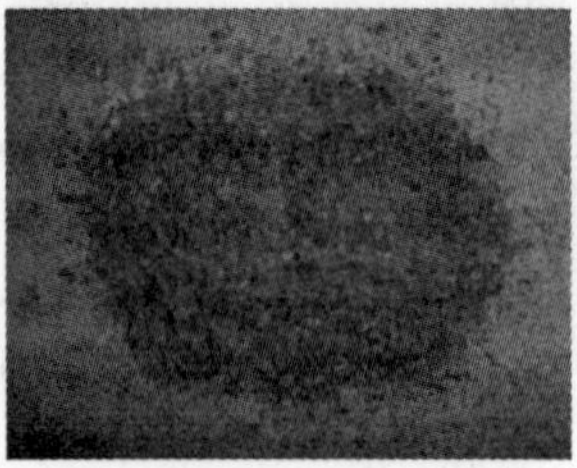

TDN	75 %
CP	11%

Other Concentrate Ingredients

Concentrate Feed Ingredients

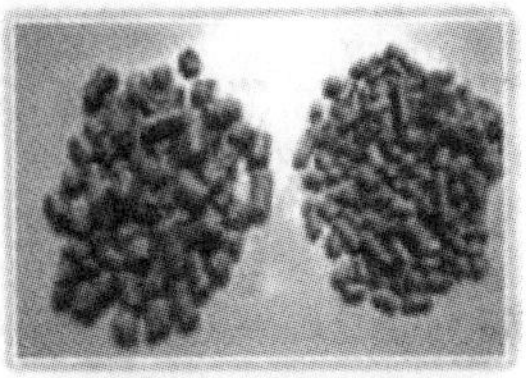
Animal Feed Pellets

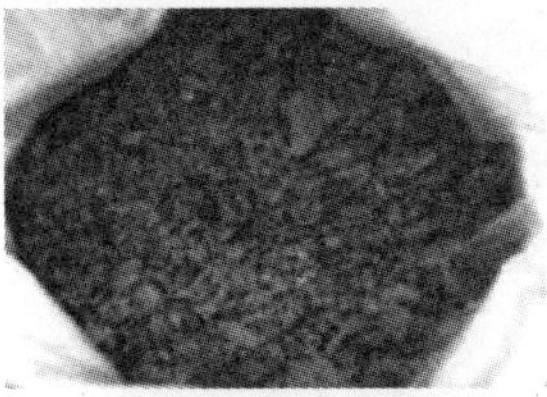
Palm Kernal Cake

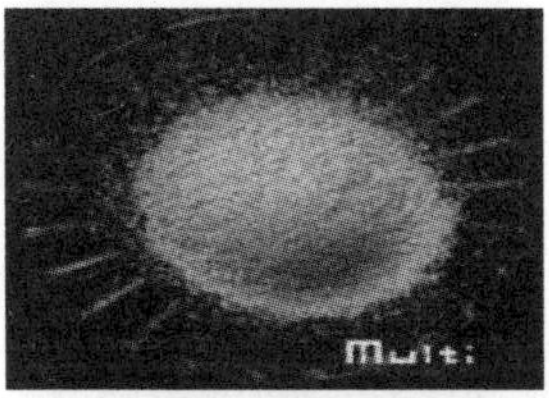
L-Lysine HCL 98%

Wheat

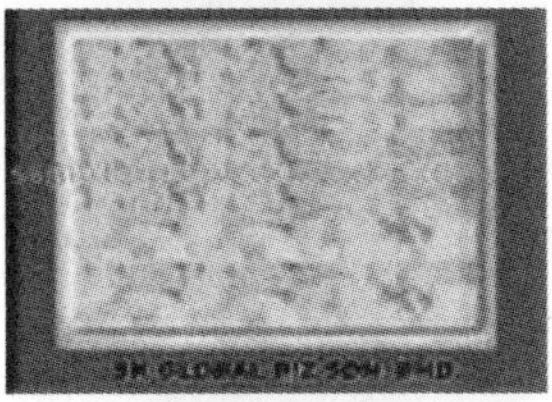
Soya Meal

Pellet

Corn meal

Cotton seed/cake

Corn Cob Meal

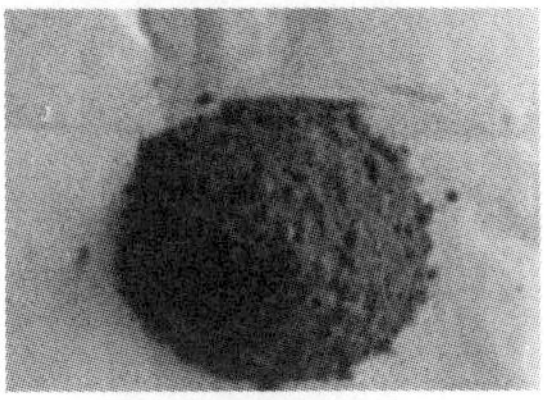
Fish Meal

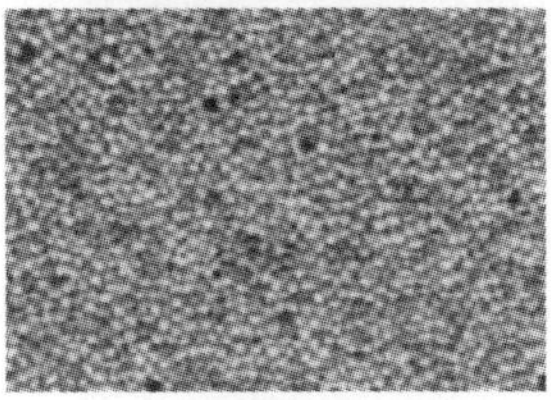
Millet

Soybean

Flaked maize

Maize gluten meal

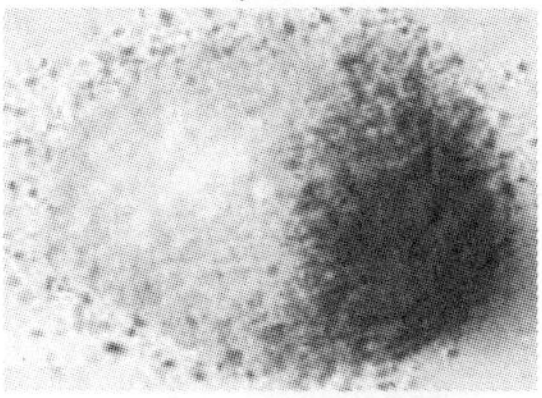
Maize gluten meal

Maize cake

Maize husk

Wheat bran

Rice bran

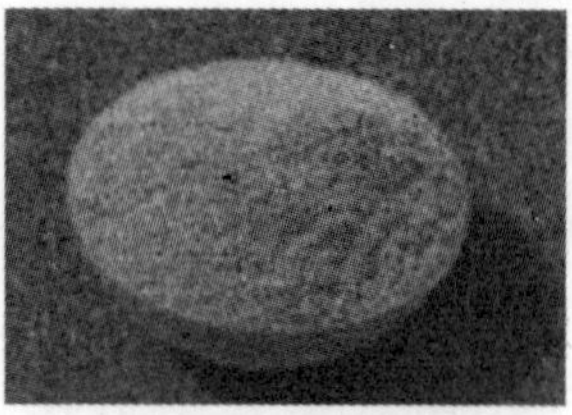
Rice polish

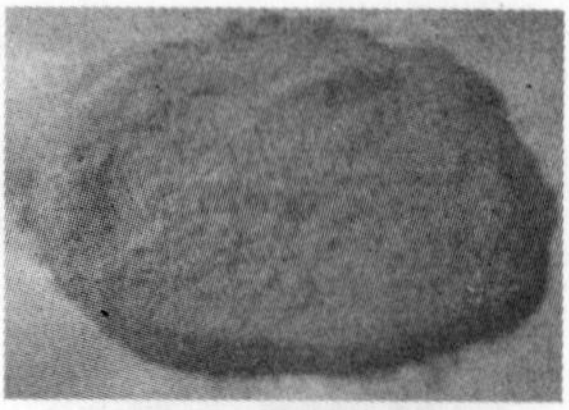
De oiled rice bran

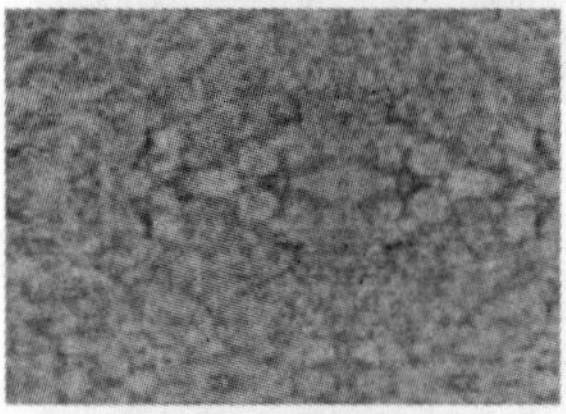
De oiled soya cake

Fish meal

Linseed cake

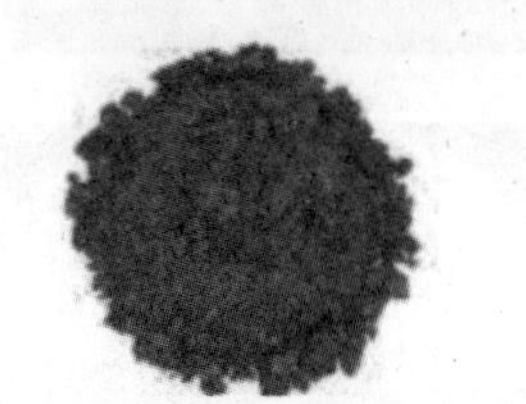
Mustard cake

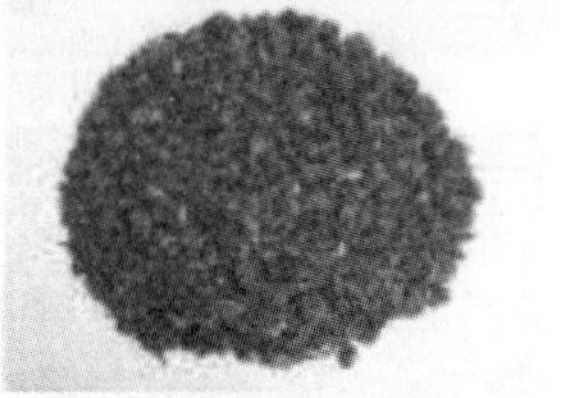
Sunflower cake

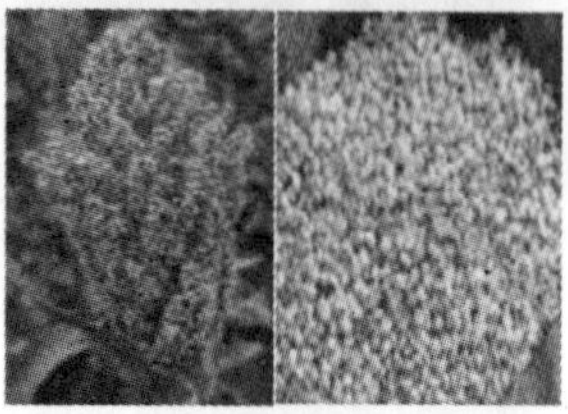
Jowar

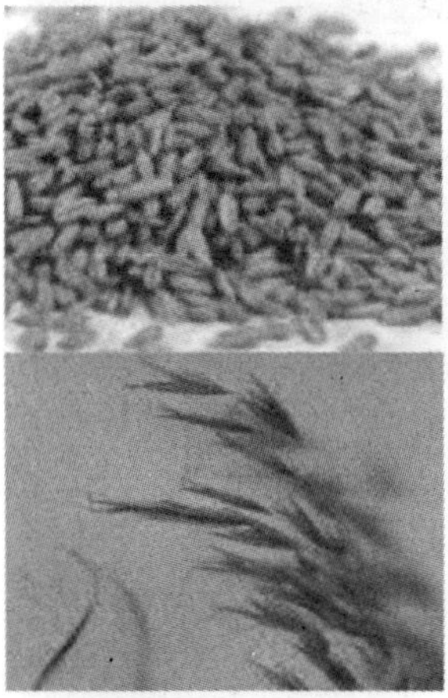
Oat

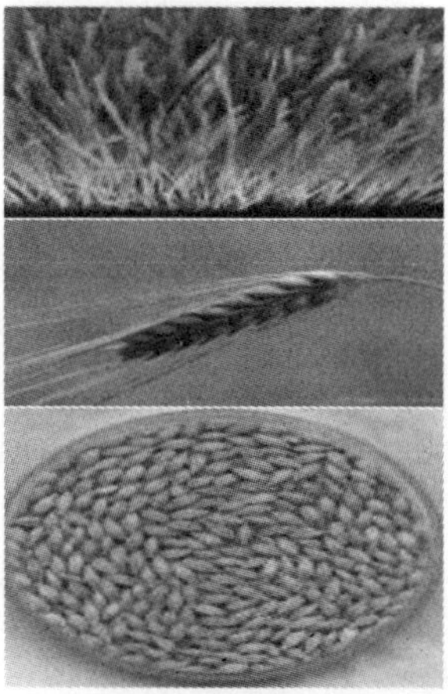
Barley

Finger millet

Bajra

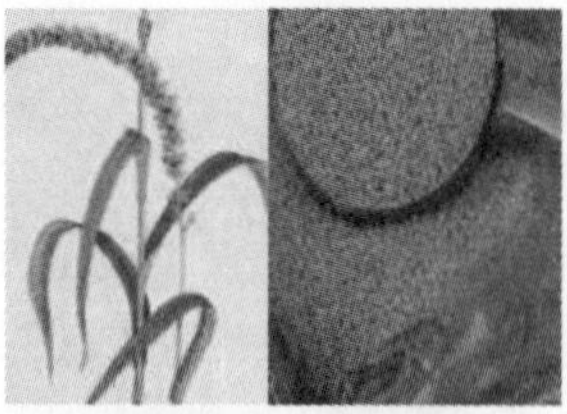
Foxtail millet

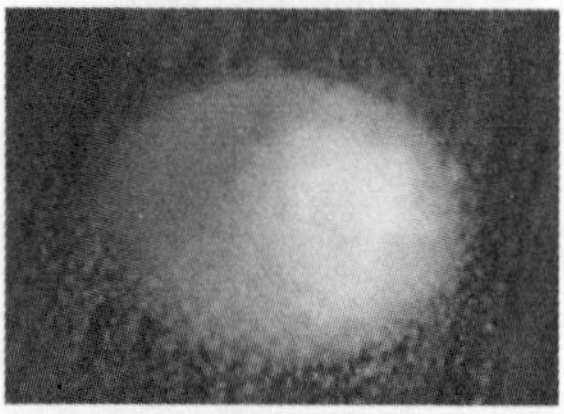
Common salt

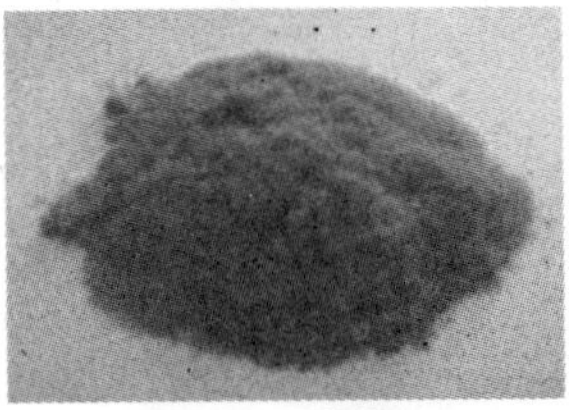
Copper sulphate

Cobalt chloride

Potassium iodide

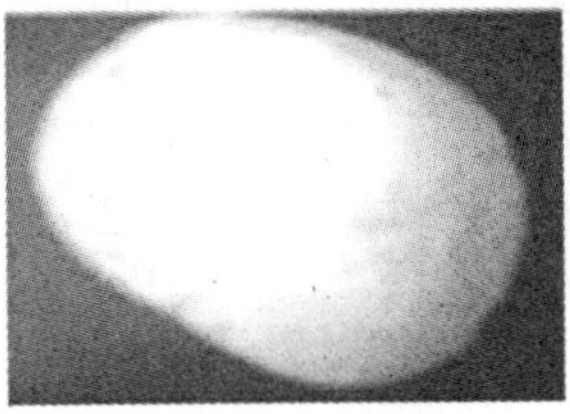
Zinc sulphate

Di calcium phosphste

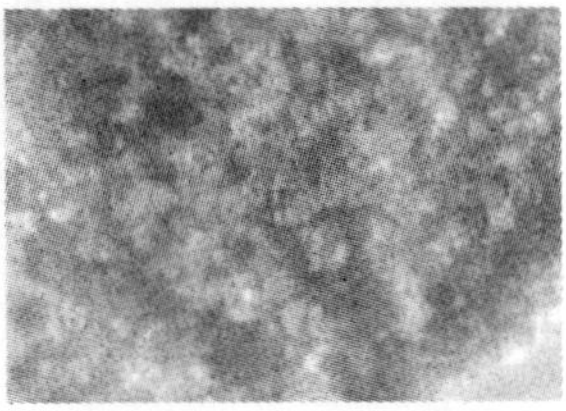
Manganese chloride

Salt block

UMMB

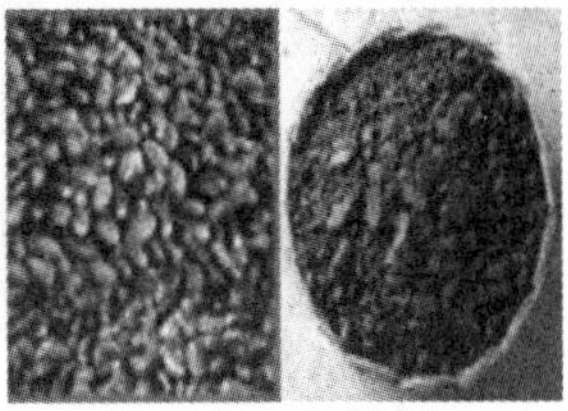
Jatropha seed/cake

Green Fodder

S.No.	Name	Feed ingredients	Nutritive value	
1.	Green maize		CP	5.0
			EE	2.7
			CF	27.4
			NFE	53.8
			TCHO	81.2
			ASH	11.1
2.	Guinea grass		CP	4.80
			EE	1.20
			CF	42.10
			NFE	39.60
			TCHO	79.70
			ASH	12.30

3. MP Chari

CP	5.30
EE	2.40
CF	32.60
NFE	52.40
TCHO	85.0 0
ASH	7.30

4. Dinanath grass

CP	5.1
EE	3.9
CF	42.7
NFE	37.8
TCHO	80.5
ASH	10.5

5. Para grass

CP	5.30
EE	2.0 0
CF	34.0
NFE	45.70
TCHO	80.30
ASH	12.40

6. Lucerne

CP	20.3
EE	3.1
CF	25.7
NFE	36.1
TCHO	61.8
ASH	14.8

7. Berseem

CP	15.50
EE	1.40
CF	17.40
NFE	53.0 0
TCHO	70.40
ASH	12.60

8. Hybrid Napier

CP	6.20
EE	2.10
CF	37.40
NFE	48.80
TCHO	82.20
ASH	9.50

□□□

Part-II

EVALUATION OF FEEDSTUFF AND FEED TECHNOLOGY

1. Determination of Proximate Principles of Feeds

During 1865 two German Scientists Henneberg and Stohman at the Weende experimental station in Germany devised a method to determine the nutrient composition of foods. This method is called the Weende method of analysis or simply proximate analysis. According to this method the nutrients present in a sample of feed are analysed and expressed in terms of six broad fundamental groups or principles called the proximate principles.

The six proximate principles are (1) Water (2) Total ash (3) Crude protein (CP) (4) Ether Extract or crude fat (EE) (5) Crude fibre (CF) and (6) Nitrogen-Free Extract (NFE).

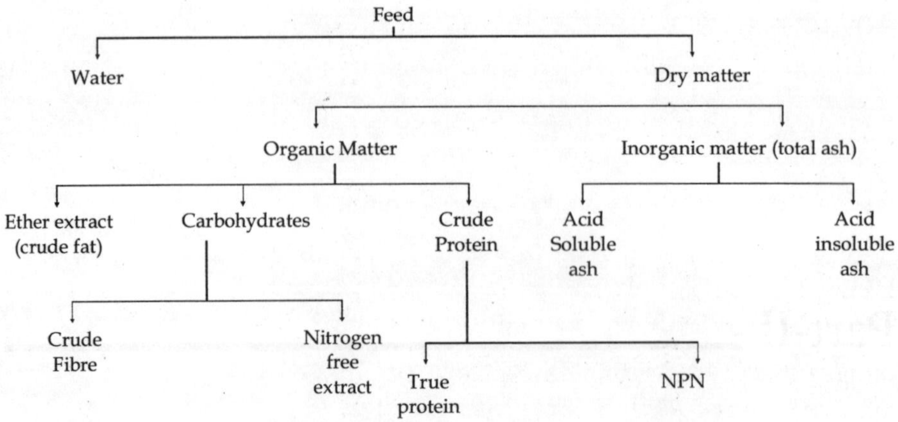

Chemical Composition *VS* Proximate Composition

In chemical composition, feed has to be analysed for all the individual chemical compounds that may be present in that feed as: glucose, starch, vit A, riboflavin, methionine, palmitic acid, fluorine etc. But such a method of analysis to determine all the individual chemical compounds present in a feed is laborious, expensive and time consuming.

The Weende system of proximate analysis of feeds is simple and yet descriptive method for evaluating the nutritive value of feeds. That is why, though this method was discovered more than 130 years ago, still it is being followed all over the world. This analysis just groups together all the nutrients present in a feed into only 6 empirical categories called proximate principles. None of the proximate principles is a chemical compound. For e.g. all the nitrogenous substances present in a feed fall into a group called crude protein; all ether solvent substances fall into one major group called ether extract. Ether extract contains not only true fats but also a variety of other substances soluble in ether. Similarly we have crude fibre which includes all the carbohydrates which are difficult to be digested and NFE, which includes all carbohydrates that are easily digestible.

Of the 6 proximate principles only 5 are actually analysed. The sixth one namely NFE is calculated by difference. The proximate principles are expressed on a percentage by weight basis, and more commonly on dry matter basis. As the NFE is calculated by difference, the cumulative error of the estimation of each of the 5 proximate principles is carried over to the NFE. Inspite of this, for all routine analysis of feedstuffs, this system of analysis is followed now.

Advantages of Weende System of Analysis

a. It is the basis for the description of foods.

b. It is a common basis for feed purchasing and for ration formulation.

c. It is the starting point for more detailed analysis.

Disadvantages of Weende System of Analysis

a. There is no direct method for estimation of nitrogen free extract.

b. Ash may contain materials of organic origin such as sulphur and phosphorus from protein.

c. There is some loss of volatile materials in the form of sodium, potassium and chlorine which will take place during ignition.

d. Crude fibre is intended originally to provide a measure of indigestible part of food but large part of it is digested by ruminants.

e. Crude fibre fraction contains cellulose, hemicellulose and lignin but a variable proportion of them will be seen in nitrogen-free extract depending on specific stage of the growth of plant material.

2. Estimation of Total Nitrogen (Crude Protein)

Principle

Protein content of food stuff is estimated indirectly by first determining the nitrogen content and then multiplying the N content by the factor 6.25, derived on the basis that all proteins contain on an average 16% of N. However for milk protein factor 6.31 is multiplied to nitrogen content of milk because its protein contains 15.84 % nitrogen. The total N present in a sample is determined by the Kjeldahl method. When a feed sample is heated with an excess of conc. H_2SO_4 in the presence of catalyst mixture the protein is hydrolysed and finally decomposed to NH_3. The NH_3 thus liberated combines with the excess H_2SO_4 present forming $(NH_4)_2SO_4$. The $(NH_4)_2SO_4$ so formed is distilled with strong alkali to liberate NH_3 which is quantitatively absorbed in a known volume and strength of standard H_2SO_4 for its final estimation and calculation of the total nitrogen present.

Apparatus

1. Kjeldahl flask 500 or 800 ml capacity
2. Distillation assembly.

Reagents

1. Conc. H_2SO_4
2. Anhydrous Na_2SO_4
3. $CuSO_4$
4. Selenium
5. 40 % NaOH
6. Std. N/10 NaOH
7. Std. N/10 H_2SO_4
8. Methyl red indicator
9. Phenolphthalein indicator
10. Zinc granules.

Procedure

I. Digestion

a. Weigh accurately a suitable quantity of the given sample on a piece of glazed paper. Transfer the material to a Kjeldahl flask.

b. Pour 25 ml of Conc. H_2SO_4 along the neck of the flask so as to wash down any particles of the sample sticking to the sides.

c. Add 10 g of anhydrous sodium sulphate or potassium sulphate, a crystal of copper sulphate and about 0.1 g of selenium.

   Note :

   - Anhydrous Na_2SO_4 or K_2SO_4 is added to raise the boiling point of the contents of the Kjeldahl flask for efficient oxidation of the material.
   - $CuSO_4$ is added as a catalyst to speed up the reaction.
   - Se is added for clarification of the digesting liquid contents in the flask.

d. Drop a few glass beads into the flask to prevent spurting while heating. Heat the flask over a Bunsen flame in a fume cupboard gently at first and then gradually to boiling point till contents become perfectly colorless. This is the end point of digestion and is reached in 2 to 3 hours of heating. When an excess of $CuSO_4$ is added the contents will be pale blue in colour.

e. When the end point of digestion is reached, remove the flame and allow the flask to cool.

Note :

- A blank experiment is conducted using all the reagents used in the actual experiment but without the sample to make a correction for the N present in the reagents used.

II. Distillation

A single unit distillation assembly consists of a 1 litre distillation flask which is connected through a trap to the upper end of an upright condenser. The trap prevents any splashing of the contents of the distillation flask into the condenser. The lower end of the condenser should almost touch the bottom of the receiver flask in which a carefully measured quantity of N/10 H_2SO_4 is taken.

1. Dilute the content in Kjeldahl flask with water rinsing the sides and neck of the flask. Rotate the flask to dissolve contents. The flask becomes very hot to touch due to the mixing of H_2SO_4 with water and evolution of heat. Allow the flask to cool.

2. Transfer the contents of Kjeldahl flask quantitatively to the distillation flask through a funnel. Wash the Kjeldahl flask several times with water and transfer the washings also. The total quantity of water used for washing should be about 200ml. Drop a few glass beads and zinc granules into the distillation flask to prevent bumping.

3. Take a measured excess quantity of standard N/10 H_2SO_4 into the receiver flask by means of a pipette or burette.

   Add 1 drop of methyl red and place the receiver flask in the distillation assembly in such a way that the tip of the condenser is well below the level of the standard acid so as to prevent any escape of NH_3 during distillation.

4. To the contents in the distillation flask add a few drops of phenolphthalein followed by an excess of 40 % NaOH slowly so that the alkali settle to the bottom without mixing with the contents of the flask. Close the flask and after assembly the distillation apparatus should be air tight.

Note :

- Ø Phenolphthalein is added to check that the contents of the flask are completely alkaline there by ensuring complete neutralization of H_2SO_4 in the the flask.

5. Heat the distillation flask and start the cold water circulation on the condenser. The distillate containing NH_3 gets collected in the receiver flask containing N/10 H_2SO_4 which absorbs all the NH_3 liberated during distillation. Distill till 150 to 200 ml of distillate is obtained in the receiver flask.
6. When the distillate has been collected, stop the water circulation in the condenser so that only steam evolves through the condenser tip. Now remove the receiver flask after rinsing the tip of the condenser with distilled water.
7. Titrate the contents in the receiver flask against standard N/10 NaOH.

Calculation

To arrive at the vol. of N/10 H_2SO_4 actually utilized, deduct the vol. of N/10 alkali required for the back titration plus the vol. of N/10 alkali required for the back titration in blank experiment from the measured vol. of N/10 H_2SO_4 taken in the receiver flask before distillation.

1ml of N/10 H_2SO_4 = 1.4 mg N

$$\text{Crude protein (\%)} = \text{No. of ml N/10 } H_2SO_4 \times 1.4 \times 6.25 \times \frac{100}{\text{Wt. of sample}} \times \frac{1}{1000}$$

Exercise

1. 1 g of groundnut cake was digested in a Kjeldahl flask. The distillate obtained was collected in 80 ml of N/10 H_2SO_4 in receiver flask. The quantity of N/10 NaOH required for back titration was 30ml. Calculate the protein content of the sample (Blank value = 0.5 ml N/10 NaOH).

2. 2 g of a layers mash was digested in a Kjeldahl flask and the digesta was distilled with alkali. The distillate was collected in 60 ml N/10 H_2SO_4 in a flask. For the back titration 12 ml of N/10 NaOH was required for the sample. If the amount of N/10 NaOH required for the blank was 1ml, calculate the protein content of the layer mash.

3. Determination of Crude Fat (Ether Extract)

Principle

When a sample of feed which is free from moisture is treated with a fat solvent like ether, the solvent extracts all the fat as well as other substances soluble in ether like sterols, lecithins, resins and volatile oils. True fat is a mixture of triglycerides whereas, the latter substances are not. Hence, the name 'Crude Fat or Ether Extract' is given for the extract obtained in this method.

Apparatus and Reagents

1. Soxhlet Extraction Assembly consists of 3 parts
   a. Condenser at the top
   b. Soxhlet or extractor in the middle and
   c. Receiver flask at the bottom. The 3 parts are assembled by means of their ground glass joints.
2. Extraction thimble
3. Petroleum ether
4. Constant temperature water bath.

Procedure

1. Weigh accurately about 5 g of the given sample (rendered moisture free) on a piece of glazed paper and transfer into an extraction thimble. The sample left in the determination of moisture preserved in a desiccator can be used for this experiment. Extraction thimbles are made of special type of fat free filter paper. When these are not available ordinary filter paper (Whatman no. 1) can be used for the purpose.
2. Take the clean, dry receiver flask from the Soxhlet assembly and weigh it accurately.
3. Introduce the thimble with sample into the Soxhlet. The top of the thimble should be well over the level of the siphon bend. Assemble the apparatus and fill Soxhlet with petroleum ether by pouring it through the condenser at the top by means of glass funnel. The amount of solvent taken is about 1.5 times the capacity of the Soxhlet.

4. Place the apparatus on a water bath at 60-80°C. Start cold water circulation in the condenser. Extract for 8h.

5. After extraction is over remove the thimble with the material from Soxhlet. Assemble the apparatus again and heat it on the water bath to recover all the ether from the receiver flask. The flask now contains only the crude fat.

6. Disconnect the receiver flask, wipe the outside of the flask thoroughly with a clean dry cloth to remove the film of moisture and dust and dry it in a hot air oven at 100°C for one hour, cool in a desiccator and weigh.

Calculation

$$\text{Ether extract (\%)} = \frac{\text{Wt. of extract}}{\text{Wt. of substance taken}} \times 100$$

Note :

- Preserve the extraction thimble with the material in a desiccator for crude fibre determination.

Exercise

1. 3 g of moisture free GN cake was extracted with petroleum ether. The weight of the receiver flask before and after extraction was 62.1850 and 62.4360 g, respectively. Calculate the fat content of the GN cake.

2. 5 g of moisture free chick mash was extracted with petroleum ether. The weights of the receiver flask before and after extraction were 52.9920 and 53.2540 g, respectively. Calculate the fat content of the chick mash.

4. Estimation of Crude Fibre

Principle

When a sample of moisture and fat free plant material is boiled first with dilute acid and then with dilute alkali the soluble carbohydrates and proteins go into solution and are extracted. The undissolved residue left behind representing cellulose, hemicellulose and lignin put together is called "Crude Fibre". When the residue after drying and weighing is heated, the fibre being organic in nature is burnt, whereas the ash or mineral matter is left over. The fibre content is obtained by deducting the weight of ash from the weight of the dried residue.

Apparatus

1. Beaker-tall form, without spout 500 ml capacity.
2. Filter flask-1000ml capacity with a rubber cork fitted with a 4" glass funnel.
3. Filter pump-glass.
4. Muslin cloth with 18 threads to a centimeter.

Reagents

1. 1.25% H_2SO_4
2. 1.25% NaOH
3. Alcohol
4. Ether

Procedure

1. Weigh an aliquot of the moisture cum fat free sample into a beaker with a mark made at 200ml level with a glass marking pencil. The moisture and fat free residue left in the extraction thimble in the previous experiment may be used.
2. Add 200 ml 1.25% H_2SO_4 and heat gently on low flame to boil for exactly 30 minutes. If the contents are boiled vigorously there will be much frothing which might be avoided.
3. Filter in a filter flask connected to a filter pump through a muslin cloth held over the glass funnel. Wash residue on cloth with boiling water till free from acid. Test with litmus paper.
4. Transfer material from muslin cloth back to the beaker. Add 1.25% NaOH up to the mark and heat to boil for exactly 30 minutes avoiding frothing.
5. Filter through the same muslin cloth on the filter flask connected to a filter pump. Wash with boiling water till free from alkali. Test with litmus paper. Wash residue with 150 ml of absolute alcohol and then with 10 ml of ether.

6. Transfer the residue to a clean dry silica crucible. Dry at 105°C in a hot air oven, cool and weigh.
7. Ignite in a muffle at 600°C for 30 minutes then cool it.

Calculation

$$\text{Crude fibre (\%)} = \frac{\text{Wt. of silica crucible with contents before ashing} - \text{Wt. of silica crucible with contents after ashing}}{\text{Wt. of substance taken}} \times 100$$

Exercise

1. 2 g of a moisture and fat free cattle feed was digested alternately with 1.25% H_2SO_4 for 30 minutes and 1.25% NaOH for 30 minutes. Calculate the crude fibre content with the following data.

   a. Weight of the silica crucible plus crude fibre residue = 33.5513g

   b. Weight of the silica crucible plus ash = 33.3200g.

2. 2 g of moisture and fat free chick mash was taken for crude fibre estimation. Calculate the crude fibre content with the following data.

   a. Weight of the silica crucible plus crude fibre residue = 32.4194g

   b. Weight of the silica crucible plus ash = 32.1706g.

5. Estimation of Nitrogen-Free Extract

Nitrogen free extract (NFE) is that fraction of carbohydrates which is obtained when the sum of CP, EE, CF, water and TA is subtracted from 100. It includes all soluble carbohydrates comprising of pentose, hexoses, disaccharides, polysaccharides, dextrins and starches. The simple sugars are readily hydrolysed from digestive tract. The starch and dextrins undergo hydrolysis by the action of digestive enzymes into glucose and absorbed. According to Weende method of analysis, the total carbohydrates are expressed in terms of CF and NFE from feed samples. It has to be estimated indirectly with the following formula.

NFE (%) = 100 - (Water + CP+EE+CF+TA)

NFE (% on DMB) = 100 - (CP + EE +CF+TA)

Exercise

1. Calculate NFE content of the given feed samples with the following data expressed on DM basis.

Crude protein	16.27
Crude fibre	7.12
Ether extract	2.59
Total ash	7.39

2. Calculate NFE content of the given feed samples with the following data

Moisture	9.17
Crude protein	7.19
Ether extract	1.73
Crude fibre	25.48
Total ash	10.91

6. Estimation of Calcium

Principle

When a solution containing calcium is treated with ammonium oxalate, all the calcium present is precipitated as calcium oxalate. The precipitate when treated with sulphuric acid dissolves, forming calcium sulphate liberating free oxalic acid which is quantitatively estimated by titration against standard N/10 potassium permanganate solution to arrive at the calcium content present in the given solution.

1 ml of N/10 $KMnO_4$ = 2 mg of calcium

Apparatus

1. 250 ml beaker with glass rod
2. Pipette 10ml
3. Funnel with stand
4. Burette with stand
5. No. 44 filter paper

Reagents

1. Dilute NH_4OH solution
2. Methyl red indicator (0.2% alcoholic solution)

3. Saturated ammonium oxalate solution
4. 20% H_2SO_4
5. Conc. H_2SO_4
6. Standard N/10 $KMnO_4$ solution.

Procedure

1. Pipette out an aliquot (10 or 20 ml) of the ash solution into a 250 ml beaker. Dilute to 80 ml with distilled water with a polythene wash bottle. Wash the sides of the beaker with little distilled water.

2. Add 10 to 15 ml of saturated ammonium oxalate solution and 2 drops of methyl red indicator. The solution turns red as the ash had been extracted with HCl in the preparation of the acid soluble ash. Methyl red has a pH range of 4.2 to 6.3. The solution which is acidic, neutralized by adding dilute NH_4OH drop so as to just reach pH 6.3 when the solution is faint pink in colour.

   Note :

   - The adjustment of pH of the solution is very important as otherwise magnesium also will get precipitated above this pH range which is an alkaline medium.

3. Boil the contents of the beaker for 10 minutes. This makes the precipitate granular.

4. Filter through No. 44 filter paper. Wash the precipitate with small quantities of hot water several times till free from oxalate.

   Note :

   - To check whether the precipitate is free from oxalate, collect half a test tube full of the filtrate, acidify with a few drops of conc. H_2SO_4 and add a drop of N/10 $KMnO_4$. Pink colour persisting shows absence of oxalate.

5. Transfer the filter paper with the precipitate carefully to the same beaker used in the experiment. Unfold it along the side of the beaker. Wash precipitate down with a jet of hot distilled water. About 50 ml of hot distilled water may be used for this purpose. Then wash with sufficient quantity of 20 % H_2SO_4, (about 20 to 40 ml) till the precipitate completely dissolves. Add 5ml of conc. H_2SO_4 if necessary.

6. Keep back the filter paper in the original funnel. Now heat the contents in the beaker to 70⁰C and titrate the contents of beaker against N/10 $KMnO_4$ to a faint pink colour. Take the filter paper and put in the beaker now. The pink colour will disappear. Add just a drop of N/10 $KMnO_4$ to get back faint pink colour. Record the volume of N/10 $KMnO_4$ required for the titration.

   (1 ml of N/10 $KMnO_4$ = 2 ml of Ca)

Calculation

$$\text{Calcium (\%)} = \text{N/10 } KMnO_4 \times 2 \times \frac{\text{Total ash solution in ml}}{\text{Aliquot of ash sol. taken}} \times \frac{100}{\text{Wt. of Sample}} \times \frac{1}{1000}$$

Sample

(To express in terms of CaO = Ca x 1.4

To express CaO to Ca = CaO x 0.714)

Exercise

1. 10 g of sesbania leaves were ashed and the ash solution was made up to 100 ml. 10 ml of the ash solution was precipated as calcium oxalate. The acidified precipitate required 14 ml of N/10 $KMnO_4$. Calculate the calcium content of the sample.
2. 10 g of fish meal was ashed and the ash solution was made up to 100 ml. 10ml of the solution was precipitated for calcium as calcium oxalate. The precipitate required 35.0 ml of N/10 $KMnO_4$ for titration. Calculate the calcium content of the sample.

7. Estimation of Phosphorus

Principle

Phosphorus is one of the chief acid forming mineral constituents of plant and animal feeds. Phosphorus present in the ash solution is precipitated as ammonium phosphomolybdate. The precipitate is washed till free from acid and dissolved in a known excess of std. N/10 NaOH. The excess alkali is back titrated against std. N/10 HCl to arrive at the exact quantity of std. N/10 NaOH.

1ml of N/10 NaOH = 0.1347 mg of P.

Apparatus

1. 250 ml beaker with a glass rod
2. Funnel with stand
3. Burette with stand
4. 5 ml pipette
5. No. 42 filter paper.

Reagents

1. Conc. HNO_3
2. Talapatra's 20% ammonium molybdate solution

   200 g of ammonium molybdate dissolved in 800 ml of distilled water in a graduated measuring cylinder. 25% NH_4OH is added until the mixture is faintly alkaline and clear. Then the volume is made up to 1 litre. The mixture is well shaken, filtered and used.
3. Phenolphthalein indicator
4. N/10 NaOH
5. N/10 HCl
6. 2% HNO_3
7. 3% KNO_3

Procedure

1. Pipette out an aliquot (5 or 10 ml) of the ash solution into 250 ml beaker.
2. Take 5 to 10 ml of conc.HNO_3 into a test tube and 5 to 10ml of ammonium molybdate reagent into another test tube and add them simultaneously into the beaker in a common stream and stir the contents with glass rod. Now the temperature of the contents of the beaker will be about 70^{0}C which is sufficient for the precipitation of the phosphorus as ammonium phosphomolybdate. Allow the precipitate to settle down over night.
3. Filter through No.42 filter paper by decantation without disturbing the ppt. Wash ppt. Only once with 2 % HNO_3 solution and then with 3 % neutral KNO_3 solution several times till free from acid.

Note :

- The ammonium phosphomolybdate precipitate is sparingly soluble in water and hence should be washed with 2 % HNO_3 and then with 3 % KNO_3 solution, but not with water.
- To check whether the precipitate is free from acid collect half test tube full of the filtrate, add a drop of Phenolphthalein and 2 or 3 drops of N/10 NaOH. Pink colour persisting shows absence of acid.

4. Transfer filter paper with precipitate to the same beaker. Add a little distilled water, stir and smash the filter paper. Then add about 10 to 20 ml distilled water.
5. Add from burette std. N/10 NaOH till all the precipitate dissolves. Now add a few drops of phenolphthalein indicator. Add 10 ml of the standard alkali in excess and note final volume.
6. Back titrate the solution with std. N/10 HCl to arrive at the actual quantity of N/10 NaOH required in the experiment.

Calculation

If total vol. of std N/10 NaOH added is = x ml

And vol. of std. N/10 HCl required for back titration is = y ml

Then vol. of std. N/10 NaOH actually required will be = (x-y) ml

1ml of N/10 NaOH = 0.1347 mg of P.

$$\text{Phosphorus (\%)} = \frac{(X-Y) \times 0.1347 \times \text{total vol. of ash solution} \times 100 \times 1}{\text{Aliquot of ash solution} \times \text{Wt. of sample} \times 1000}$$

(to express in terms of P_2O_5 = P x 2.29

to convert P_2O_5 to P = P_2O_5 x 0.436)

Exercise

1. 5 g of a feed was ashed and the ash solution was made upto 100 ml. 5 ml of this solution was precipitated as ammonium phosphomolybdate. This precipitate was dissolved in 60 ml N/10 NaOH. If the volume of N/10 HCl required for back titration was 5 ml, calculate the phosphorus content of the sample.

2. 2 g of a concentrate feed was ashed and the ash solution was made upto 100 ml. 5 ml of this solution was precipitated as ammonium phosphomolybdate. This precipitate was dissolved in 50 ml N/10 NaOH. If the volume of N/10 HCl required for back titration was 15 ml, calculate the phosphorus content of the sample.

8. Determination of Neutral Detergent Fibre

Though proximate analysis system is the starting point of chemical analysis, yet it has certain limitations. The major limitations lie with crude fibre (CF) and nitrogen free extract (NFE). As per proximate analysis, CF should contain total cellulose, hemicellulose and lignin. However, only part of the hemicellulose and lignin comes out in the crude fibre fraction. The remaining portion shows up as NFE, which is usually thought consisting mainly of highly digestible sugars and starch. Consequently, the hemicellulose low in digestibility and lignin essentially indigestible appears in NFE fraction, thereby giving an incorrect index regarding nutritive value of a given feed.

To overcome these limitations, Van Soest (1963 and 1967) developed a new scheme of analysis of forage dry matter in terms of cell wall and cell constituents using various detergents. According to this scheme, cell contents, which are soluble in neutral detergents, include lipids, sugars, organic acids, starch, soluble proteins, non protein nitrogenous compounds and other water soluble matters. While the insoluble material known as neutral detergent fibre (NDF) comprises cell wall constituents which include hemicelluloses, some cell wall nitrogenous compounds and acid soluble ash pass into solution leaving all other cell wall constituents in the insoluble fraction which is named as acid detergent fibre (ADF) or lingo cellulose. This lingo-cellulose when treated with 72 % H_2SO_4, the cellulose is dissolved in the acid leaving behind the lignin and acid insoluble ash, from which lignin can be determined by loss of ignition.

Schematic Representation of Van Soest Method of Analysis

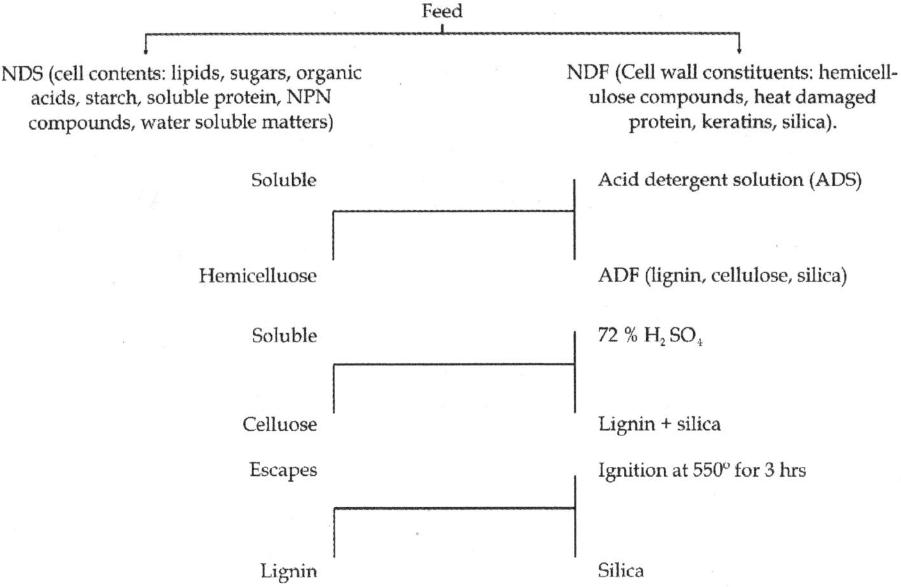

Advantages

1. Fractionation of fibre into hemicelluloses, cellulose and lignin is possible.
2. Prediction of digestibility and nutritive value of feed stuffs.

Disadvantages

Cellulose may not be soluble completely and this may result in increased percentage of lignin.

Determination of Cell Contents and Cell Wall Constituents

Apparatus

1. Refluxing apparatus (as in CF estimation)
   a. Tall pyrex or corning beakers (spoutless) of about one litre capacity.
   b. Round bottom flask as condenser.
2. Sintered glass crucibles with coarse porosity (grade I) of about 50 ml capacity.
3. Other equipments used for CF estimation may be adapted in these procedures.

Reagents

1. Neutral detergent solution

   a. Composition

   Distilled water = 1 litre,

   Sodium lauryl sulphate = 30 g

   Disodium ethylene diamine tetra acetate (EDTA) dihydrate crystals (reagent grade) = 18.61 g

   Sodium borate decahydrate (Borax, reagent grade) = 6.81 g

   Disodium hydrogen phosphate anhydrous (reagent grade) = 4.36 g

   Ethoxy ethanol (ethylene glycol monoethyl ether) purified grade = 10.0 ml

   b. Preparation

   Put EDTA and sodium borate decahydrate ($Na_2B_4O_7$. 10 H_2O) in a large beaker (2 litre). Add some of the distilled water, shake and heat until these are dissolved. Now add sodium lauryl sulphate and 2 –ethoxy ethanol. Put disodium hydrogen phosphate (Na_2HPO_4) in a separate beaker and add some of the distilled water and heat until it is dissolved. Mix both the solutions properly and check pH to range between 6.9 and 7.1. If made properly, pH adjustment will rarely be required.

2. Decahydronaphthalene (Dekalin, reagent grade)
3. Acetone: use grade which is free from colour and leaves no residue on evaporation.
4. Sodium sulphite, anhydrous (reagent grade).

Procedure

1. Weigh 0.5-1.0 g air dry sample (ground to pass through 20-30 mesh screen) into a beaker of the refluxing apparatus.
2. Add 100ml neutral detergent solution, 2.0 ml decahydronaphthalene and 0.5 g sodium sulphite.
3. Filter through a weighed glass crucible on filter manifold. Rinse the sample into the crucible with minimum of hot (90-100^{0}C) water. Filter liquid and repeat the washing procedure.
4. Wash twice with acetone in the same manner and suck dry.

5. Dry the crucible in hot air oven at 100°C for 8 h or overnight and weigh.
6. Ash the residue in the crucible at 500-550°C for 3h and weigh. Record ash content as ash insoluble in neutral detergent solution.

Calculation

1. Cell wall

$$\text{Cell wall constituents (\%)} = \frac{(\text{Wt. of crucible + Cell wall constituents}) - \text{Wt. of crucible}}{\text{Wt. of sample}} \times 100$$

2. Cell contents (%) = 100 – cell wall constituents

9. Determination of Acid Detergent Fibre, Cellulose, Hemicellulose and Lignin

Determination of Acid Detergent Fibre (ADF)

Apparatus

Same as used in NDF estimation.

Reagents

1. Acid detergent solution

   a. Acid detergent solution

   i. Reagent grade H_2SO_4 standardized to = 49.04 g/l (100 % assay)

   ii. Cetyl trimethyl ammonium bromide (CTAB) = 20 g.

   b. Preparation

   Add 20 g of CTAB to 1 N H_2SO_4 and make the volume to one litre

2. Decahydronaphthalene (Dekalin, reagent grade)
3. Acetone as used for NDF estimation
4. n-hexane (technical grade)

Procedure

1. Weigh 1g air dried sample (ground to pass 20-30 mesh screen) into a beaker of the refluxing apparatus.
2. Add 100ml acid detergent solution and 2 ml decahydronaphthalene.
3. Heat to boiling in 5 to 10 minutes. Reduce heat as boiling begins, in order to avoid foaming. Reflux for 60 minutes, timed from onset of boiling.
4. Filter through a weighed glass crucible on filter manifold. Rinse the sample into the crucible with minimum of hot water (90-100°C). Filter the liquid and repeat washing procedure.
5. Wash twice with acetone in the same manner and suck dry. Break up all lumps so that the solvent may come with all particles of fibre.
6. Wash with hexane, if required. Hexane should be added while crucible still contains some acetone. (Hexane use can be avoided if lumping is not a problem).
7. Suck the acid detergent fibre free from hexane and dry the crucible at 100°C for 8h or overnight in hot air oven and weigh.

Calculation

$$\text{Acid detergent fibre (\%)} = \frac{\text{(Wt. of crucible + fibre) – Wt.of crucible}}{\text{Wt. of sample}} \times 100$$

Acid Detergent Lignin (ADL)

Estimation of ADL requires preparation of ADF first.

Apparatus

Same as required in NDF estimation, muffle furnace.

Reagent

H_2SO_4, (72% w/w)

Preparation

Take 417ml distilled water in a volumetric flask and add 583 ml pure H_2SO_4 slowly with occasional swirling.

(caution: Flask must be cooled in a water bath while mixing H_2SO_4)

Procedure

1. Prepare the ADF
2. Fill the crucible (sintered) containing ADF with 72 % H_2SO_4 and stir with glass rod to smooth the paste and break the lumps. Let glass rod remain in the crucible; refill with 72 % H_2SO_4 and stir at hourly intervals as acid drains away. Crucible does not need to be kept full at all times. Three additions would be sufficient. Keep the crucible at 20-23^{0}C.
3. After three hours, filter off as much acid as possible. Wash the contents with hot water until it is free from acid. Rinse and remove glass rod.
4. Dry the crucible at 100^{0}C for 8 h or overnight and weigh.
5. Keep the crucible in muffle furnace at 500-550^{0}C for 3 h. Cool it and weigh.

Calculation

$$\text{Cellulose (\%)} = \frac{\text{Wt.of crucible before treating with 72 \% } H_2SO_4 - \text{Wt.of crucible after treating with 72 \% } H_2SO_4}{\text{Wt. of sample}} \times 100$$

$$\text{Acid detergent Lignin (\%)} = \frac{\text{Wt. of crucible before ignition} - \text{Wt.of crucible after after ignition}}{\text{Wt. of sample}} \times 100$$

$$\text{Silica (\%)} = \frac{\text{Wt. of crucible before ignition} - \text{Wt. of empty crucible after ignition}}{\text{Wt. of sample}} \times 100$$

Hemi cellulose (%) = % NDF - % ADF

Exercise

1. a. Calculate NDF content of the feed sample with the following data: Wt. of sample 0.5929 g; wt of empty crucible 63.7348 g and wt. of crucible + NDF residue 64.2144 g.

   b. Calculate the different fibre fractions (ADF, hemi cellulose, cellulose, lignin and silica) of the given feed with the following data. Wt. of sample 0.6055 g, wt. of crucible 57.32 g, wt. of crucible

+ ADF residue 57.6245 g, wt. of crucible + residue after treating with 72 % H_2SO_4 57.3850 g, wt. of crucible + ash 57.332 g.

2. a. Calculate NDF content of the feed sample with the following data: Wt. of sample 0.4427 g, wt. of crucible 63.745 g, wt. of crucible + NDF residue 64.0788 g.

   b. Calculate ADF, hemi cellulose, cellulose, lignin and silica content of the given feed with the following data. Wt. of sample 0.4112 g, wt. of crucible 57.3315 g, wt of crucible + ADF residue = 57.5498 g, Wt. of crucible + residue after treating with 72% H_2SO_4 = 57.3636 g, wt. of crucible + ash = 57.3482 g.

10. Quality Control of Feedstuffs

1. Physical Tests to Identify Common Adultrants of Feed

Physical inspection (visual inspection) will give a good assessment of quality of feed ingredients.

Physical Inspection

(1) colour (2) odour (3) taste (4) evidence of wetting (5) evidence of deleterious substances or foreign materials (6) storage pests and (7) evidence of damaged grain. First 3 items can be done with reference to normal and abnormal by experience of having normal things in mind. Eg. maize, jowar, wheat, rice etc.

Evidence of wetting: Moisture level less than 13% are considered as safe for storage. Any grain generally will have below 10% moisture. Even upto 13 % we can store safely without mould infestation. More than 13% moisture creates hot spots which causes mould growth.

Evidence of deleterious substances like dust, sand, straw, clay, nails etc. can be find out by visual appearance.

Storage pests: We can see the insects and faeces and eggs of insects by visual inspection and by smell especially in wheat bran, rice bran etc. If we suspect, we can test and observe the insect eggs under microscope.

Evidence of damaged grain: Damaged or broken grain is unsafe for storage. Broken material will absorb more moisture and sometimes we see insects. In unbroken grain, seed coat is intact and will be resistant for moisture absorption.

The physical inspection is important for routine quality control and it will be effective only by experience.

2. Feed Microscopy

Stereoscopic microscope- one having two optical systems which give depth to the object being viewed or the appearance of solidity – will be used for microscopic examination of feed.

a) Low magnification – particle size, softness, hardness etc.

b) High magnification – requires more skill, accurate results are obtained.

Note :

- Microscopist should have at hand, collection of pure feed ingredients, adulterants, contaminants. After working with known materials one should proceed to identify the ingredients in simple mixture, subsequently complete mixtures.

The Use of Feed Microscopy in Quality Control

The use of microscopy in feed quality control can give even the smallest feed manufacturer some measures of protecting against adulteration and contamination of ingredients. Products may be examined for the presence or absence of basic ingredients and by means of spot tests or tracer ingredients which give much information on the presence of drugs or other micro-ingredients.

Procedure for the Qualitative Estimation of Ingredients

1. *Sample preparation:* Reduce sample by quartering if meal, can be examined without further preparation. If pellets, reduce by crushing with pestle and mortar. If animal products or fish products or mixture with high fat content, extract at least 200 g with ether. If facilities are available, determine moisture, ash, protein, fat, fibre, Ca, P and NaCl.

2. *Separation:* Separate the portions by using different sieves and place portions obtained in glass dishes or other containers suitable for use under the microscope. Weigh each portion, these weights may be of aid in determining relative amounts of certain ingredients. If necessary the mineral portion may be separated from 1.0 g portion by floatation with chloroform or carbon tetrachloride.

3. *Examination :* For determining of single ingredient or impurity examine each portion, starting with the portion contained on the largest screen and examine by going through the portion using a desecting needle or fine pointed forceps to single out the item concerned. It may be necessary to resort to flotation with chloroform to separate the light and heavy particles.

   Combine the identified particles selected from the several portions in a container. Weigh and calculate percentage. This procedure will be found useful for meat scraps, meat and bone scraps, blood meal, fish meal and similar products.

4. Identification aids (by colour of particles)

   White-Limestone, dicalcium phosphate, dried milk, bentonite etc.

   Tan or Buff- Soybean meal, Vit A. beads, bone meal etc.

   Reddish or brown – blood meal, meat scraps, maize bran, fish meal, linseed meal etc.

   Black – blood meal, meat scrap, charcoal, fish meal

   Yellow – yellow maize meal, cotton seed meal etc.

5. Identification aids (by treatment with reagents)

   These tests are best applied by placing about two drops of the reagent in a clear or white spot plate or filter paper and sprinkling a small portion of the feed. They may also be used to identify single particles.

Reagent	Reaction	Compound indicated
HCl 0.5N Ammonium molybdate	Effervescence	Carbonates
Solution	No ppt. Yellow ppt	Calcium carbonate Bone meal, dicalcium phosphate
Silver Nitrate 0.1 N	White ppt. No ppt.	Salt No salt
Distilled water	White solution	Milk products
Conc. H_2SO_4	Effervescence blue colour Light red Red purple Orange brown	Carbonates or NaCl Chlorate Chlortetracycline Oxytetracycline, Tetracycline Riboflavin

Microscopic Characters of Some Feed Ingredients

Ground maize – Yellow, entire grain ground, maize bran and traces of maize cob.

Groundnut cake – Light to dark tan, hulls, skins and characteristic odour.

Soybean meal – Light to medium yellow, hulls pitted usually curled helium oval and thick.

Cotton seed meal – Brownish yellow – hulls are thick, curled dark brown. Characteristic twisted cotton fibres and hull fragments, very light yellow to dark brown in colour depending on the method, temperature, and length of processing.

3. Spot Tests for Minerals, Antibiotics and Urea

Qualitative colour tests which indicate the presence or absence of certain additives in feeds are useful as quality control tools to production and laboratory personnel, to regulatory officials and to others concerned with the safe and effective use of feed additives.

i. Urea

Not considered as a drug and is most certainly on important additive for ruminant feeds as a source of NPN. However, since ingredient is considered contaminant in non-ruminant feeds, a simple test paper method is given for detection of urea in feeds i.e. Urease-Bromothymol blue test paper for urea.

a. Urease solution – Wet 0.2g urease powder with small amount of water, stir into paste and dilute to 10ml with water.

b. Bromothymol blue solution – Rub 0.15 g indicator powder in mortar with 2.4 ml 0.1 N NaOH solution. After indicator dissolves, wash mortar and pestle with distilled water and dilute to 50ml with water. Solution should be green, pH approximately 7.

c. Test paper – Mix 10ml of indicator solution (b) with 10ml urease solution (a). Pour mixture into watch glass using clean tweezers, dip pieces of heavy filter paper (Whatman No. 5) in solution (to avoid uneven distribution of indicator and enzyme, wet entire paper at one time by laying in on surface of solution). Hang paper to dry in place free from NH_3 fumes, strong air currents or heat.

Paper should be orange when dry. Store dry paper in well stoppered dark glass bottle in cool place.

Procedure - Moisten urea test paper with few drops of distilled water. Sprinkle feed evenly over the paper. A dark blue colour will develop if urea is present.

ii. Test for HCN

Cyanogenic glycosides in feeds

Prepare Na - picrate paper by dipping strips into 1% picric acid solution and drying. Then dip into 10% Na_2CO_3 solution and dry. Preserve these papers in stoppered bottles.

Finely chop small quantity of plant material and place in a test tube. Insert a piece of the moist Na- picrate paper in tube taking care that it does not come in contact with material. Add few drops of chloroform and stopper tube tightly. The Na-picrate paper gradually turns orange and then brick red if plant tissue contains cyanogenic glycosides. The test is delicate and rapidity of change in colour depends on quantity of free HCN present.

iii. Chlortetracycline and Oxytetracycline

Trade name of oxytetracycline is Terramycin while that of chlortetracycline is aureomycin.

Reagent

Sakaguchi - modified

Dissolve 10 g crystalline boric acid in 300ml distilled water and carefully add while stirring, 700 ml of conc. H_2SO_4.

Transfer to a glass stoppered bottle. Better results may be obtained if stored in refrigerator and used cold. Reagent is stable for several months.

Procedure

Place a small amount of feed sample in a white spot plate. Treat with 2 or 3 drops of Sakaguchi reagent. Chlortetracycline will instantly form a deep purplish blue colour. Oxytetracycline (Terramycin) gives brilliant red colour instantly when the reagent is added.

iv. Furazolidine & Nitrofurazone

Reagents

1. 4% potassium hydroxide- (4 g KOH dissolved in 96 ml of denatured 95% ethyl alcohol).
2. Dimethylformamide
3. Mix 10 ml of KOH solution with 90ml dimethyl formamide. This solution should be freshly prepared before use.

Procedure

Place about 25ml of reagent 3 in a 50 ml beaker, take small quantity of the feed to be examined on a spatula. Hold it just above the beaker, tap to release small amount of feed at a time, observe the solution while particles are descending.

Furazolidine particles produce blue trials as they are discending. Nitrofurazone particles produce red trials as they descend. Carryout the examination in good light at eye level for best results.

v. Amproleum

Solution A: 1.25 g potassium ferricyanide + 3 g NaOH + 500 ml water.

Solution B: 50 ml 27 naphthalene diol + 500 ml methyl alcohol+150 ml water.

Solution C: Extraction solvent. (Equal parts of water, methyl alcohol and chloroform).

Procedure

1. Add 4ml of solution A to 13ml of solution B and mix.
2. Add about 10 g of feed to be tested to 30 ml of extraction solution C. Shake for one minute.
3. Let the feed settle for 1 minute.
4. Decant approximately 5 ml of the supernatant into combine A and B solutions and mix.
5. If amproleum is present, the colour should start to change in seconds. The colour reaches its maximum intensity in 1 to 3 minutes with the formation of purple coloured solution.

vi. Nitrates and Nitrites

Applicable to grain, forage and other feedstuffs.

Reagents

1. Dissolve 0.5 g of diphenylamine in 20 ml of dist. water. Cautiously add conc. H_2SO_4 to make 100ml of solution.
2. Conc. H_2SO_4.

Procedure

Place the material to be tested on a white background such as a porcelain spot or china cup. Some materials such as hay may be moistened with a few drops of water prior to the test if necessary. Using equal parts of reagent 1 and reagent 2, add the reagent drop wise to the material being tested.

If nitrate is present a blue colour develops. The intensity of the colour and the speed of development indicate the approximate amount of nitrate present. If nitrite is present a greenish colour develops. Nitrite is 10-15 times more toxic than nitrate.

vii. Test for Salt, Lime stone or Phosphate

To a teaspoonful of feed in a test tube add 5-10 ml of chloroform and shake. Float the organic matter off. Then repeat the washing of mineral residue with more chloroform. Shake the residue out of the tube on to a clean paper and evaporate the chloroform leaving the dry materials. Taste portion of this dry residue to note the addition of salt. Limestone and phosphate will impart a gritting feeling.

viii. Decomposition Test

Used for animal and marine products. Into a 250 ml conical flask put 5g of sample to be tested. Prepare a cork, which fits tightly with a 2″ x ¼″ strip of white paper pinned to the bottom and moistened with saturated solution of lead acetate. Add the sample in the flask 50ml of dilute H_2SO_4 prepared by addition of 5ml of conc. H_2SO_4 to 45ml distilled water. Insert cork and let stand in warm room for 16hour. If the sample is badly decomposed, the test paper will darken quickly. This test does not definitely determine toxicity but if a sample causes intense darkening, it is not satisfactory for feeding.

11. Silage Preparation

Silage is the material produced by the controlled fermentation of a crop of high moisture content.

Crops for Silage Making

The crops at the time of ensiling should have 35% D.M. The crops should be rich in sugars for the fermentation. The crop should have solid stem so that small amount of air is trapped. However, hallow stemmed crops can also produce good silage but trampling should be adequate. Maize is a popular crop for silage. It possesses the required DM % and available sugars. Thus normal fermentation is insured without the addition of any preservatives. Fodder sorghum is better for silage than the grain sorghum. Grasses and legumes when used for silage are usually referred to as hay crop silage. Specialized methods required for preparing good hay crop silage are 1) addition of molasses, 2) wilting of fodder before ensiling to reduce moisture content and 3) proper trampling of fodder at the time of ensiling in the case of hallow stemmed fodders.

Type of Silos

a. Based on structure
   i. Cylindrical or tower silos
   ii. Trench or bunker silos
   iii. Pit silos

b. Based on the material used for construction
   i. Brick silo
   ii. Cement silo
   iii. Stainless steel silo
   iv. Katcha silo. The silo where none of the above construction material is used is termed katcha silo or pit silo.

Cylindrical silo

Bunker silo

Horizontal bunker silo

Bunker silo

Bag silo

Drive-over pile

Trench silo

Galvanised silo

Katcha Silo

Katcha silos are prepared by digging a pit or trench in the ground. The sides and floor of silo are generally plastered with a mixture of cow dung and clay (1:1) for making the walls and floor smooth. The bottom and sides should be carpeted with dry grass of kadbi (chopped) or long straw of grasses, cereal crops etc. so as to make a 5-6 cm thick carpet all round. It prevents the direct contact between the freshly chopped material and the soil. The direct contact between the chopped material and the soil decreases or increases the moisture content in the ensiled material, depending upon the soil type and water table. Another way is to spread black polythene film of 0.2 mm thickness to cover the floor and sides of the pit and the remaining film has to be covered over the ensiled material before closing the pit.

Steps for Making Silage

1. The fodder crops should be harvested at the stage of growth so that the moisture, protein and carbohydrate contents are optimum for anaerobic fermentation.

2. The fodder to be ensiled should be chopped into small pieces preferably of size 1-2 cm using a chaff cutter. Finely chopped forage packs more densely, it makes better silage and the loss of nutrients is reduced to a minimum. Also the silage can be removed much easier if the forage is chopped.

3. The silo pit should be filled very quickly (within 3-4 days) and the material must be completed in such a way so as to remove as much air as possible through constant pressing either by manual labour or with bullocks or tractor or combination of all the three. The safest plan to ensure good silage is to have a reliable man in the silo as it is filled, to distribute the forage uniformly. By walking about, he will also help pack the forage.
4. The level of chopped material should be about 1-2 meters, above the ground level. During the course of fermentation, the material will gradually settle down.
5. The silo pit, after filling and compacting the material carefully, should preferably be given a dome like shape. Such shape facilitate the drainage of rain water, which otherwise would enter the ensiled material and deteriorate the quality.
6. The ensiled material is then covered with a polythene sheet from all sides. However, if the polythene sheet is not available a thick layer of straw is put on the chopped material from all sides and over the straw, thick layer of moist soil (10-12 cm) is spread. The surface is covered in a mud plaster prepared by adding dung and water to the earth in such a way that the ratio of clay and dung is 1:1.
7. The chopped material ensiled by the above procedure is ready for feeding to the livestock after 2-3 months. A silo pit is opened and the material is removed daily for feeding by exposing as little surface of the ensiled material as possible. This is done mainly to prevent the direct contact of the feeding material with sun-light and prevent air drying.

Proper Size of Silo

The diameter of silo should depend on the number and kind of animals to be fed from it, and its height on the length of the feeding period. The silo should be of such diameter that 5 cm of silage will be removed from the entire surface daily to prevent spoilage. Knowing the no. of animals to be fed, the quantity of silage required daily may be ascertained. 5 cm in depth of ordinary corn silage average about 38 kg/m^2 in a silo filled to a depth of 9 minutes.

To use 5 cm daily from the surface approximately the following amounts will be obtained from silos of various diameters.

Diameter		Weight
3mt	-	280 kg
3.6mt	-	410 kg
4.2mt	-	550 kg
4.8mt	-	720 kg
5.4mt	-	910 kg
6mt	-	1230 kg

When the maximum diameter which the silo should have has thus been determined, the total amount of silage required for the desired period may be completed and the dimensions for the silo of this capacity found by referring to the table. It should be born in mind that silage in relatively deep silos keeps better than in shallow one, as has been pointed out before. Also when a silo is more than 5.5 or 6 mt in diameter, more labour is required in carrying the silage to the silo door and it is removed for feeding.

Weight of Silage and Capacities of Silos

Silage is much more compact and heavier as the depth from the surface increases. The weight for m3 at any depth will vary considerably, depending especially on the % water in the crop when ensiled and the amount of grain in it.

The following table shows the estimated capacity of silos of various sizes for well-eared corn silage ensiled in the early dent stage, cut in one-quarter inch length, well trampled during filling of the silo, and with the silo refilled once after settling for one day. The depth indicated is the actual depth of the settled silage not the height of the silo.

Table : Tonnes of settled maize silage in cylindrical silos.

Inside diameter (meter)	Depth of the silage in meter											
	1.2	1.8	2.4	3.0	3.6	4.2	4.8	5.4	6.0	6.6	7.2	7.8
3.0	5	8	11	15	19	23	27	31	35	39	43	47
3.6	7	11	16	22	27	33	38	44	50	56	62	68
4.2	9	15	22	29	37	44	52	60	68	76	85	93
4.8	12	20	29	38	48	58	68	79	89	100	110	121
5.4	15	25	37	49	61	74	86	100	113	126	140	151
6.0	18	31	45	60	75	90	107	123	139	156	173	190
	8.4	9.0	9.6	10.2	10.8	11.4	12.0	12.6	13.2	13.8	14.4	15.0
3.0	52	56	60	64								
3.6	74	80	87	93								
4.2	101	110	118	126	135	143	152					
4.8	132	143	154	165	176	187	198	209	220			
5.4	162	181	195	209	223	237	251	265	279	203	307	320
6.0	206	224	241	258	275	292	310	327	344	361	378	396

Estimation of Amount of Silage in a Silo

To estimate the amount of silage remaining in silo after part has been fed out, find the actual depth left and estimate the original total depth of silage after settling. Then compute the amount as follows.

Let us suppose that 3 mt of silage is left in a silo having a diameter of 4.2 mt and that after settling the entire depth of the silage was about 8.4 mt. before feeding started. From the table we find out that at first the silo contained 101 tones. The first 5.4 mt. of silage which was fed out, contained about 60 tones, according to the table. The difference or 41 tones is approximate weight of the silage left.

12. Laboratory Ensiling of Green Fodders

To assess the quality of silage from a new variety of fodder in the laboratory, small quantity of fodder can be ensiled in two ways.

1. Silage Bags

These are made of smaller tubes of plastic. They may be small silos of 50 kg or even less. These are useful for small livestock operations and these are inexpensive. The ends of each tube are double tied making a bag with handle at each end.

2. Plastic Containers

Plastic buckets with lid or smaller plastic containers with lid can also be used as silos. The chopped material is tightly packed in these containers and capped tightly.

13. Feed Mixing

Feed preparation involves 4 steps

a. Formulation

b. Proportioning or batching

c. Grinding

d. Mixing or blending

Ration is defined as the nutrients required for animals for 24 h for maintenance, production, growth etc. Nutrients should be in proper proportion and should pass through the gut in that period.

a. *Formulation* – Ration formulation is done by 3 methods

i. Pearson square method

ii. Trial and error method

iii. Weighted averages

b. *Batching* – A formula before it is mixed has to be batched or proportioned and weighed. Batching is nothing but weighing of all the ingredients as per formula of that batch.

Scales – Scales are nothing but weighing machines which are usually used for weighing the ingredients as per formula. Scales may be semi-automatic or automatic. If the scales do not act accurately, the formula will not be proper and it will be costing customer or the owner. So the ingredients should be weighed accurately.

After batching, the ingredients will be kept on the intake platform. Intake platform means the place from which the ingredients will go to their respective places i.e. maize, groundnut cake etc. which require grinding, to the hammer mill; brans, which do not require grinding, to mixer and micro ingredients for preparing premix.

Premix : Premix will be about 2% of the formula. The micro ingredients weighed accurately in small balance, then mixed with some inert substance like bran and then transferred into mixer. Micro ingredients should not be sent to mixer directly.

Vertical Feed mixer

Horizontal feed mixer

Feed grinder

Feed pelletiser

Feed Extruder

c. *Grinding* - Grinding is the first and important processing method in feed processing. It is a process of particle size reduction. Grinding may be coarse, medium and fine according to the resultant particle size. Grinding of feedstuffs, concentrates and roughages are usually done by hammer mills. Principle involved in hammer mill is impaction. Feed ingredients are reduced into smaller particles by means of hammers and impact surface. Component parts of hammer mill are

1. Feeder
2. Dome
3. Shaft
4. Hammers
5. Rotor-contains several plates keyed to the shaft
6. Sieve or screen
7. Pulley or shaft
8. Motor with pulley
9. Belt of motor
10. Exit

Hammer mills are commonly used in feed industry at farm and commercial levels.

Working

Material is fed through the feeder while the motor is running. Due to the movement of the hammers at high speed (2100 to 7500 mt/min) impaction develops. The material will be in free flight and touches the hammer tips or screen and gets reduced to smaller size. If the first impact of the hammers against the grain does not break it up so that it will fall through the screen, it rebounds and is again struck by the hammer tips. The process continues until all particles are reduced to a size that will allow them to pass through screen and comes out through exit.

d. *Mixing or blending* - Mixing is defined as blending of 2 or more different types of ingredients. Mixing is important of all the processes in feed processing operation and hence the name feed mixing plant.

There are two types of mixers which are commonly used in feed industry:

a. Horizontal mixers

b. Vertical mixers

a. Horizontal Mixer

Components parts are

1. Hopper - for holding the ground material
2. Small hopper - for introduction of mealy ingredients
3. Small opening - For adding micro-ingredients
4. Window opening in the hopper to observe the proceedings
5. Horizontal mixer- has a round bottom and square top- A trough is present on the horizontal mixer which has small openings for molasses to pass through it. Inside the mixer, ribbons or paddles are moulded on center driven shaft.
6. Discharge gate-located at the center of the bottom.
7. Motor with belt and speed reducer
8. Legs or stand on which the mixer is kept
9. Platform

Working

The ground materials are introduced one by one through the small hopper. Start the motor. If the formula contains molasses, preheat the molasses and introduce the quantity during mixing at the last phase. After complete mixing the material is discharged through discharge gate. Surge bin holds the compounded material for faster bagging or binning of the material. 3-5 minutes are required for completing the operation.

Advantages

1. Molasses can be mixed
2. 100% clean out
3. Short mixing time (3-7 minutes)
4. Surge bins can be used for continuous mixing.

Disadvantages

1. Consumes more power
2. Can not be used by small mills due to high cost.

b. Vertical Mixer

Component parts are

1. Screw conveyor
2. Window
3. Feeder
4. Exit
5. Legs
6. Motor
7. Pulley
8. Belt.

Working

Introduce the ground mealy ingredients. Dumping of material should be in the order of bulk ingredients, minor ingredients and micro ingredients. Close the inlet after introduction of all ingredients. Start the motor. The feed moves with the help of screw and it will be dropped on either side. By repeated elevation of material from bottom to top results in blending. This has to be run for 15-20mts depending on formula. After mixing, the material is taken through the exit. Stop machine after complete discharge of material from mixer.

Advantages

1. Low power consumption.
2. Can be used in small feed plants due to low capital investment.

Disadvantages

1. Long mixing time (20-30 mts).
2. Liquid supplements more than 2-3% cannot be mixed.
3. Clean out is not 100%.
4. Surge bins cannot be used for continuous mixing.

❑❑❑

SECTION-II
Part-I

LIVESTOCK FEEDING

1. Some Important Terms and Definitions

Nutrient: The chemical substances found in feed materials which are necessary for the maintenance, production and health of animals.

Feed (Feedstuff): Food of animals comprising any naturally occurring ingredient or material fed to animals for the purpose of sustaining growth and development.

Diet: A regulated selection of a feed ingredient or a mixture of ingredients including water, which is consumed by animals on a prescribed schedule. A balanced diet supplies all nutrients needed for normal health and productive functions.

Ration: A fixed amount of feed for one animal, fed for a definite period of 24 hr.

Balanced Ration: The ration which provides proper amount, proportion and variety of all the required nutrients to keep the animal in its form to perform best in respect of production and health.

Nutrient Requirements : Gives the requirements for optimum production and amounts of nutrients that must be supplied when fed in a ration to meet the needs of a normal healthy animal in hygienic environment.

Nutrient Allowance: Gives an extra allowance of nutrient over the requirement as a margin of safety.

Maintenance: Amount of feed that is needed just to support an animal when doing no work and yielding no product.

When an animal receives only maintenance ration, its body will neither gain nor lose in protein, fat or mineral matter.

AAFCO- Association of American Feed Control Officials.

Additive: An ingredient or combination of ingredients added to the basic feed mix or parts thereof to fulfill a specific need. Usually used in micro quantities and requires careful handling and mixing.

Blending (process) - To mingle or combine two or more ingredients of feed. It does not imply a uniformity of dispersion.

Blocks (Physical form)- Agglomerated feed compressed into a solid mass cohesive enough to hold its form and weighing more than 1 kg and generally weighing 15-20kg.

Bricks (Physical form)- Agglomerated feed, other than pellets, compressed into a solid mass cohesive enough to hold its form and weighing less than 1 kg.

Byproduct (part) - Secondary products produced in addition to the principal product.

Cake (physical form) - The mass resulting from the pressing of seeds, meat or fish in order to remove oils, fats or other liquids.

Carriers- An edible material to which ingredients are added to facilitate uniform incorporation of the later into feeds. The active particles are absorbed, impregnated or coated into or onto the edible material in such a way as to physically carry the active ingredient.

Chaff (Part) - Glumes, husks or other seed coverings together with other plant parts separated from seed in threshing or processing.

Chopped, chopping (Process) - Reduced in particle size by cutting with knives or other edged instruments.

CLFMA- Compound Livestock Feed Manufacturers Association (India).

Complete feed- A nutritionally adequate feed for animals other than man, by specific formula is compounded to be fed as the sole ration and is capable of maintaining life and/or promoting production without any additional substance being consumed except water.

Cooked, cooking (Process) - Heated in the presence of moisture to alter chemical and /or physical characteristics or to sterilize.

Cracked, cracking (process) - Particle size reduced by a combined breaking and crushing action.

Crumbled, crumbling (process) - Pellets reduced to granular form and offered to poultry, fish etc.

Dehulled, dehulling (process) - Having removed the outer covering from grains and other seeds.

Dry-rendering (process) - Residues of animal tissues cooked in open steam jacketed vessels until the water has evaporated. Fat is removed by draining and pressing the solid residue.

Expanded, expanding (process) – Subjected to moisture, pressure and temperature to gelatinize the starch portion. When extruded, its volume is increased due to abrupt reduction in pressure.

Extracted, mechanical (process) – Having removed fat or oil from materials by heat and mechanical pressure (expeller).

Extracted, solvent (process) – Having removed fat or oil from materials by organic solvents.

Extruded (process) – A process by which feed has been pressed, pushed or protruded through orifices under pressure.

Feed grade- Suitable for animal consumption.

Flakes (Physical form) - An ingredients rolled or cut into flat pieces with or without prior steam conditioning.

Flour (part) – Soft, finely ground and bolted meal obtained from the milling of cereal grains, other seeds or products. It consists essentially of the starch and gluten of the endosperm.

Free choice- A feeding system by which animals are given unlimited access to the separate components or groups of components constituting the diet.

Gelatinized, gelatinizing (process) – Having the starch granules completely ruptured by a combination of moisture, heat and pressure, and, in some instances, by mechanical shear.

Germ (part) – The embryo found in seeds and frequently separated from the bran and starch endosperm during the milling.

Gluten (part) - The tough, viscid nitrogenous substance remaining when the flour or wheat or other grain is washed to remove the starch.

Groats (part) - Grain from which the hulls have been removed.

Ground, grinding (Process) – Reduced in particle size by impact, shering or attrition.

Heads (part) – The seed or grain-containing portions of a plant.

Kernel (part) – Awhite grain. For other species dehulled seed.

Kibbled, kibbling (process) – Cracked or crushed baked dough or extruded feed that has been cooked prior to or during the extrusion process.

Mash (physical form) – A mixture of ingredients in meal form.

Meal (Physical form) - An ingredient which has been ground or otherwise reduced in particle size.

Micro-ingredients – Vitamins, minerals, antibiotics, drugs and other materials normally required in small amounts and measured in milligrams, microgram's or parts per million (ppm).

2. Digestion and Metabolism Trials in Large and Small Ruminants

Nnutrients present in the feedstuffs are not completely available to the animal body. Large portion of the nutrients are excreted in the faeces because of being not digested in the alimentary tract. Therefore, digestibility of the feedstuffs is defined as that portion of a feed or any of single nutrients of feed which is not recovered in the faeces i.e. the portion which has been absorbed by the animal. When the digestibility is expressed in percentage it is known as digestibility coefficient.

For example, if a bullock is eating 40kg of jowar green fodder containing 10kg of dry matter per day and is excreting 4kg of dry matter in the faeces then 6kg of dry matter is digested every day. When this digestibility of dry matter is expressed in percentage (6/10 x 100 = 60 per cent), then it becomes digestibility coefficient. Digestibility coefficients are calculated for dry matter, crude protein, crude fibre, ether extract and nitrogen free extract. For ash, digestibility coefficient is not calculated since most of the minerals of endogenous origin are re excreted in to the gut.

Measurement of Digestibility

The digestibilities of feeds and their nutrients are determined by conducting digestion or metabolism trial. The complete experiment requires a group of similar animals of a species, feed(s) to be evaluated, stall for individual feeding and collection of faeces, faeces collection bags and metallic cages designed for individual collection of urine and faeces in an uncontaminated form.

Material Required

1. *Experimental animals:* Adult animals of same sex, age and comparable body weight are required. Six animals will be ideal for conducting digestion trials.
2. *Animal weighing balance:* 1000 kg capacity weighing scale is required to weigh the animals before and after conducting the experiment with large ruminants and 100 kg balance for small ruminant species.
3. Metabolism stall with facilities for individual feeding and watering and quantitative collection of faeces and urine. Urine collection is required in metabolism trials and balance trials.
4. Weighing balances of 1, 5 and 25 kg capacity for weighing feeds fodders and faeces.
5. Buckets or tubs with cover for the collection of faeces.
6. Graduated buckets for recording water intake and measuring cylinders to measure urine output.
7. Plastic bottles of 2.5L capacity or buckets of 20L capacity are required for small or large ruminants respectively, for urine collection in metabolism stalls.
8. Polythene sample bags are required to store aliquoted faeces samples in deep freeze.
9. Metallic cages of appropriate size to be used.
10. Laboratory facilities for proximate analyses to be created.

Procedure

1. Adequate amount of test feed is to be collected randomly and stored in labeled polythene bags properly for further analysis. In case of fodders, regular supply at a particular stage of plant growth to be ensured.
2. *Preliminary period* : The test feed is to be fed either as sole diet or component of a composite diet. It depends on the nature of feeds. Fodders are mostly fed alone. During this period the animals are to be fed on the fodder/feedstuffs in question for a period of at least 21 days to remove the effect of previous feed and for adoption. This period is followed by a collection period during which daily record of feed intake and faeces voided is maintained on a 24 hr basis.
3. *Collection period* : Since feed intake and faeces voided show day to day variation, the collection period needs to be extended to several days. Usually 7 to 10 days collection is taken in the case of ruminants and other herbivores. During this period representative samples of feed offered, residue left and faeces voided are to be taken daily and dried, pooled and stored for further analysis.
4. *Body weight recording* : The experimental animals are to be weighed consecutively for 3 days at a fixed time in the morning before feeding and watering. The average of 3 days body weight has to be considered as the initial body weight. The same is repeated at the end of collection period.
5. a. *Recording of feed intake* : The time of offering of feed is fixed during the preliminary period at 8.00 am. Weighed amount of feed is offered at a fixed time and animals are allowed to eat during 24 hrs. A representative sample is simultaneously preserved in the laboratory for the determination of dry matter. Sample of fresh feed is processed for the determination of nitrogen.

   b. Weight of residual feed is recorded on the next day at 8 am and processed in a similar manner as in the case of feed.
6. *Recording of faeces voided* : Faeces voided during 24hrs is collected quantitatively and processed for preserving aliquots for nitrogen estimation and dry matter determination. The aliquot of faeces has to be decided on the basis of daily faecal void and may range from 1/100 to 1/1000 part for nitrogen analysis and stored in a deep freezer in labeled polythene sample bags. At the end of

collection period these daily pooled samples are defrost and aliquoted for nitrogen analysis.

7. *Recording of urine excreted :* Urine excreted during 24hrs period should be collected quantitatively and aliquoted at the rate of 1/100 part into a kjeldahl flask containing conc. sulphuric acid for nitrogen determination. Similar quantity is to be taken in a silica basin and dried for mineral analysis.

Recording of Data

Similar data sheet can be prepared for determining digestibility coefficient of organic nutrients of dry matter.

Table 1: Initial body weight (kg) data recording of experimental animals.

Day	Animal 1	Animal 2	Animal 3	Animal 4	Animal 5	Animal 6
1st day	295	320	310	330	300	320
2nd day	305	315	310	325	300	325
3rd day	300	315	315	325	305	320
Mean Wt.	300	317	312	327	302	322

Table 2 : Record of green fodder intake (Animal No. 1)

Day	Fodder offered (kg)	Residue left (kg)	Feed intake (kg)
1st	35	7	28
2nd day	30	3	27
3rd day	30	1	29
4th day	30	2	28
5th day	30	1	29
6th day	30	3	27
7th day	30	2	27
Average	30	2	28

Table 3 : Record of DM estimation in feed (Animal No. 1)

Day	Wt. of tray(g)	Wt. of fodder(g)	Total wt. (g)	Dried wt. (g)	DM (g)	DM%
1st day	100	50	150	112	12	24
2nd day	100	50	150	111	11	22
3rd day	100	50	150	114	14	28
4th day	100	50	150	113	13	26
5th day	100	50	150	112	12	24
6th day	100	50	150	114	14	28
7th day	100	50	150	113	13	26
Average						25

Table 4 : Record of DM determination in feed residue (Animal No. 1)

Day	Wt. of tray(g)	Wt. of residue (g)	Total wt. (g)	Dried wt. (g)	DM (g)	DM%
1st day	120	50	170	134	14	28
2nd day	120	50	170	133	13	26
3rd day	120	50	170	135	15	30
4th day	120	50	170	134	14	28
5th day	120	50	170	133	13	26
6th day	120	50	170	135	15	30
7th day	120	50	170	134	14	28
Average						28

Table 5 : Record of DM estimation in feed (Animal No. 1)

Day	Faeces voided (kg)	Wt. of tray(g)	Aliquot for DM (g) (1/100)	Total wt. (g)	Dried wt. (g)	DM (g)	DM%
1st day	4.5	120	45	165	145	25	55.6
2nd day	4.5	120	45	165	147	27	60.0
3rd day	4.4	120	44	164	146	26	59.0
4th day	4.8	120	48	168	149	29	60.4
5th day	4.7	120	47	167	148	28	59.6
6th day	4.6	120	46	166	147	27	58.7
7th day	4.8	120	48	168	149	29	60.4
Average	4.6						60.0

Table 6 : Final body weight (kg) data recording

Day	Animal 1	Animal 2	Animal 3	Animal 4	Animal 5	Animal 6
1st day	305	320	310	330	310	325
2nd day	300	316	315	328	305	320
3rd day	301	315	310	326	305	321
Mean Wt.	302	317	312	328	307	322

A model data sheet for tabulation and computation of dry matter digestibility

Date/Day	Weight of			Dry matter in			Total dry matter				Digestibility Coefficient (%)
	Fresh feed offered (kg)	Feed left as residue (kg)	Faeces voided in 24h	Feed offered (%)	Residue left (%)	Faeces voided (%)	Offered (kg)	Residue (kg)	Intake (kg)	Faeces (kg)	
1.											
2.											
3.											
4.											
5.											
6.											
7.											

• Similar data sheet can be prepared for determining digestibility coefficient of organic nutrients of dry matter.

Calculations

1. Body weight of experimental animals during digestion trial

Animal No.	Body weights of experimental animals (kg)			
	Initial	Terminal	Mean	$W^{0.75}_{kg}$
1	300	302	301	72.26
2	317	317	317	75.13
3	312	312	312	74.24
4	327	328	328	77.07
5	300	306	303	72.62
6	320	320	320	75.66

At 300 kg body weight average DM intake = $\frac{28 \times 25}{100}$ = 7.00 kg per day

or

$$\frac{7.0 \times 100}{301} = 2.33 \text{ kg/100 kg b.wt. / day}$$

or

$$\frac{2300}{72.26} = 32.25 \text{ g/kg } W^{0.75}\text{/day}$$

Digestibility Coefficient of Dry Matter (DM)

DM offered / day = 7.70 kg

DM left/day = 0.78 kg

DM intake/day = 7.70-0.78 = 6.92 kg

DM voided in faeces /day = DM in aliquot x 100

= 27 x 100g = 2700g or 2.70 kg

DM digested = DM intake – DM voided in faeces

= (6.92-2.70) kg

= 4.22kg

$$\text{Digestibility coefficient of DM} = \frac{\text{DM digested} \times 100}{\text{DM intake}}$$

$$= (4.22 \times 100)/6.92$$

$$= 60.98 \text{ (or) } 61\%$$

Determination of Digestibility Coefficients of other Organic Nutrients

The composite samples of feed offered, residue left and faeces voided should be analysed for proximate principles and other nutrients (if digestibility coefficients has to be determined). However, to carryout nitrogen analysis for crude protein determination wet samples of faeces preserved in frozen condition should be taken up. The digestibility coefficient of each organic nutrient of proximate principles has to be calculated as per the procedure followed for DM digestibility.

Determination of Balance of Nutrients

Using the nutrition studies balances of carbon, nitrogen, minerals and energy is normally determined. For the determination of balances of nitrogen and minerals, a metabolism trial is to be conducted on the experimental animals. The procedure is exactly similar to that of digestion trial except that the quantitative collection of urine is also necessary. For the determination of the balances of minerals, intake of water also needs to be recorded and analysed along with the mineral under study in the feed stuff.

Quantitative Collection of Urine

The urine is to be collected in a receptacle containing either toluene or sulphuric acid. These are used to stop the loss of nitrogen due to enzymatic changes. For the collection of urine, collection funnels are to be harnessed during the metabolism trial in specially designed cages. Painted metallic cages are used for the collection of urine when the balance of metals is determined.

Calculation

N-Balance (g/day) = N intake (g) – {loss of N in faeces (g) + loss of N in urine (g)}

Eg: A calf weighing 100 kg has consumed 72.6 g of nitrogen per day and excreted 28.6 and 22.0 g of faecal N and urinary N, respectively. Calculate N balance of the calf.

N balance = 72.6 – (28.6+22)
= 72.6 – 50.6
= 22.0 g

Q: Calculate the N balance of the calf when it consumed 75g of nitrogen and excreted 35.5 and 18g of faecal and urinary N, respectively.

Ans: N balance = 75-(35.5 + 18)
= 75-53.5
= 21.5 g

Exercises

1. A buffalo has consumed 30kg of green maize having a dry matter content of 25% and the animal excreted 2.5kg of DM. Calculate the amount of DM digested and the digestibility coefficient of DM.

2. Six bullocks on average consumed 20kg of green fodder per head per day containing 25% DM and 1.6% CP and voided 20kg of faeces containing 10% DM and 0.6% CP per day. Calculate the digestibility coefficients of DM and CP of the green fodder under test.

3. Calculate the digestibility coefficients of DM and organic nutrients of a test feedstuff using the following data:

Total quantity of feed offered to the animal	=	40 kg
DM of feed offered	=	25 %
Total refusal	=	6 kg
DM of the refusal feed	=	30%
Total faeces voided	=	4 kg
DM of the faeces	=	50%

Proximate composition (on DMB) of feed offered and faeces voided are:

Proximate principle	Feed %	Faeces (%)
CP	8.0	10.0
EE	2.0	3.5
CF	37.0	28.0
NFE	40.0	36.5
Total ash	13.0	22.0

3. Calculation of Digestible Crude Protein (DCP) and Total Digestible Nutrients (TDN)

The digestibility coefficients of crude protein (CP), ether extract (EE), crude fibre (CF) and nitrogen-free extract (NFE) are to be calculated. The value of digestible EE is multiplied with 2.25 and added to digestible CP, CF and NFE values to get the TDN values of a particular feedstuff. The energy equivalent of EE is about 9 against 4 for carbohydrates and protein.

DCP = Dig. Coefficient x CP

TDN (%) = % dig. CP + (% dig. EE x 2.25) + % dig. CF + % dig. NFE

Eg: Calculate the TDN content of paddy straw using the following data:

Dig. CP = 0.3 kg; Dig. EE = 0.5kg; Dig. NFE = 18.9kg; dig. CF = 19.7 kg

TDN = 0.3 + 18.9 + 19.7 + (2.25 x 0.5)

= 40%

Exercises

1. Calculate Total Digestible Nutrient (TDN) content of cotton seed from the following data generated in an experiment.

Proximate principle	Composition %	Digestibility coefficient
CP	18.2	72
EE	18.4	85
CF	25.0	63
NFE	27.5	60

2. Calculate the nutritive value (DCP and TDN) of the feed from the following data:

Proximate principle	Composition %	Digestibility coefficient
CP	8.0	74
EE	2.0	47
CF	37.0	78
NFE	40.0	75

3. Calculate the nutritive value of the feed mentioned in Sl no. 3 of previous exercise by using the data given in the problem.
4. Read and record the CP, DCP and TDN values of locally available forages and feedstuffs that are commonly fed to livestock and poultry.

4. Indirect Method of Determining the Digestibility or Digestibility by Difference

The digestibility coefficients of individual feeds/forages may be determined if it provides a satisfactory ration for the period of test when fed alone like, green maize, jowar, grasses, legumes etc. Some feeds, however, cannot be fed alone and therefore their digestibility coefficients cannot be determined directly like concentrates such as cakes, grains and brans as they do not supply the bulk. In case of non-maintenance type roughages like straws and kadbies also, directly digestibility determination cannot be made since during the test period the feed stuff in question do not supply the required nutrients.

In this process, two or more digestibility trials need to be conducted. In the first trial the digestibility of nutrients has to be determined by feeding a basal maintenance type fodder where the test feed in question is not fed. For instance, if the digestibility of nutrients of groundnut cake is to be estimated, first trial has to be conducted by feeding green maize fodder. In this trial digestibility of nutrients present in the maize green fodder will be determined. In the second trial maize green fodder and groundnut cake will be fed together. In this case it is assumed that the digestibility coefficients of the nutrients in the maize fodder will remain the same in the second trial as was determined in the first trial. Therefore, from the total undigested nutrients voided in the faeces the portion belonging to the maize fodder is subtracted. The remaining faecal nutrients are considered to have come from the groundnut cake or the feedstuff under test.

Example : From the first trial it was found that maize fodder had 60 per cent digestibility for dry matter and 62.5% for crude protein. In the second trial 20kg of green maize and 1 kg of groundnut cake were fed to the animal. The animal excreted on an average 22kg of faeces. The following was the composition of green maize, groundnut cake and faeces.

	Green maize	Groundnut cake	Faeces
%DM	25	90	10
%CP	1.6	45	0.7

$$\text{Total DM intake} = \frac{25 \times 20}{100} + \frac{1 \times 90}{100} = 5.9\ \text{kg}$$

$$\text{Total DM voided in faeces} = \frac{10 \times 22}{100} = 2.2\ \text{kg}$$

DM voided in the faeces from the GNC = 2.2 – 2.0 = 0.2 kg

DM digestibility of cake = 0.9 – 0.2 = 0.7 kg

$$\text{Digestibility coefficient of DM of GNC} = \frac{0.7}{0.9} \times 100 = 77.7\%$$

In a similar way the CP digestibility of groundnut cake can be calculated.

Exercise

1. Calculate the digestibility coefficient of CP of groundnut cake from the above data.

5. Indicator Method of Determining Digestibility

Often conducting of conventional digestion trial is laborious and time consuming. Many investigators have sought many indirect methods for assessing the digestibility. It involves the use of inert reference substance as an indicator. Chromic oxide has been found to be very useful for this purpose as an external marker. Lignin and plant chromogens can also be used as (internal) indicators since these are the natural constituents of feedstuffs. These markers could be quantitatively determined with the help of spectrophotometer by determining the ratio of concentration of the reference substance to that of a given nutrient in the feed and the same ratio in the faeces resulting from that feed, the digestibility of nutrients can be obtained without measuring either the feed intake or faecal output using the formula:

$$\text{Digestibility Coefficient} = 100 - \left[100 \times \frac{\text{\% of indicator in feed}}{\text{\% of indicator in faeces}} \times \frac{\text{\% of nutrient in faeces}}{\text{\% of nutrient in feed}}\right]$$

Exercises

1. The per cent of Chromic oxide is 0.8 in feed and 1 in faeces. The per cent of CP in feed is 8 and in faeces 2. Calculate the digestibility coefficient of CP of the feed.

2. The per cent of Chromic oxide is 0.8 in feed and 1 in faeces. The per cent of CF in feed is 60 and in faeces 35. Calculate the digestibility coefficient of CF of the feed.

6. Measurement of Pasture Consumption and Digestibility

The problem of measuring how much feed an animal obtains from a pasture and determining the digestibility of feed are very much important factors for feeding range cattle, buffaloes and sheep. Various methods have been proposed for determining the feed intake.

1. One of the methods is that animals are harnessed with faeces bags. The DM consumption can be determined by collecting the the total faeces in the faecal bags by which the DM voided can be determined. Simultaneously the DM digestibility of the fodder is determined by feeding the fodder to the animals in metabolism stalls.

   For eg: An animal was voiding 2,500 g faeces daily. The digestibility of DM determined separately was 50% that means the animal consumed 5,000 g of DM through the pasture.

2. DM intake can also be determined by the indicator. This indicator can be an internal or an external like chromic oxide. There is no need of harnessing the animal with faecal bags, since the faecal output can be determined by external indicator and the internal indicator will indicate the DM digestibility to determine the pasture consumption as shown in example 1 of exercise.

$$\text{DM consumption (g/d)} = \frac{\text{Units of internal indicator Per g of dry faeces} \times \text{g of DM in faeces per day}}{\text{Units of internal indicator per g DM of forage}}$$

Estimation of Total Faeces by Grab Samples

This is done when the collection bags cannot be used. In such cases a grazing animal is fed a measured amount of an external indicator and grab samples (these are small samples of faeces removed from the rectum manually or collected at random from the pasture at various times during the day) of faeces subsequently collected. The total faecal DM output can be calculated as indicated in the following equation:

$$\text{Faecal DM output (g)} = \frac{\text{External indicator fed (g)}}{\text{\% external indicator in faeces grab sample DM}} \times 100$$

Note: Grab samples should be taken from each animal to keep error below 5 % and the chromic oxide should be fed at least 10 days, before the start of sampling.

Exercise

1. Calculate the DM intake and protein digestibility of an animal maintained on range conditions by using the following data.

   a. Fed 10g Chromic oxide per day

b. Feed contained	-	13% Lignin	(DMB)
	-	7.7% Protein	(DMB)
c. Faeces contained	-	1.15% Chromic oxide	(DMB)
		22.6% Lignin	(DMB)
		6.0% Protein	(DMB)

$$\text{Faeces output (g DM)} = \frac{10}{1.15} \times 100 = 870 \text{ g of faecal DM}$$

$$\text{i. DM intake} = \frac{870 \times 22.6}{13} = 1512 \text{ g}$$

$$\text{Apparent digestibility Coefficient of protein} = 100 - \left[100 \times \frac{13}{22.6} \times \frac{6}{7.7}\right] = 55.2\%$$

$$\text{For DM} = 100 - \left[100 \times \frac{13}{22.6} \times \frac{100}{100}\right] = 42.5\%$$

Indigestible DM = 100 – 42.5

= 57.5%

$$\text{ii. DM consumption} = \frac{\text{DM in faeces (g)}}{\text{\% indigestibility of DM}} \times 100$$

$$= \frac{870 \times 100}{57.5} = 1512\text{ g}$$

2. Calculate the DM intake and protein digestibility in an animal maintained on range conditions by using the following data.

   a. Fed — 10g Chromic oxide per day

   b. Feed contained 12% Lignin (DMB0, 8 % Protein (DMB)

   c. Faeces contained - 1.1% Chromic oxide (DMB), 20.0% Lignin (DMB), and 6.0% Protein (DMB).

7. Calculation of Requirements of Nutrients

(In terms of DCP and TDN for Maintenance Growth and Other Production Traits)

The feeding standards that are currently followed in India are recommended by ICAR (1985) for maintenance, growth and other production traits in various species of domestic livestock.

Nutrient requirements of cattle and buffaloes.

1. Maintenance

For calculating maintenance energy needs of cattle and buffaloes, the requirements base has been taken as 142 Kcal of ME/Kg W 0.75. The value varies from 1.84 to 2.60 g/kg W0.75 for DCP. The data are shown in the following table.

Table 1: Nutrient requirements for maintenance of cattle and buffaloes

Live weight (Kg)	DM (Kg)	DCP (g)	TDN (Kg)	CA (g)	P (g)
250	4-5	140	2.20	25	17
300	5-6	168	2.65	25	17
350	6-7	195	3.10	25	17
400	7-8	223	3.55	28	20
450	8-9	250	4.00	31	23
500	9-10	278	4.45	31	23
550	10-11	310	4.90	31	23
600	11-12	336	5.35	31	23

2. Milk Production

Extra requirements of energy, protein and minerals for production of milk have to be added to the maintenance of such cows as presented in Table 2.

Table 2: Nutrient requirements per kg of milk production based on milk fat

Fat (%)	DCP(g)	TDN(g)
3.0	48	275
3.5	51	300
4.0	55	325
4.5	58	350
5.0	62	375
5.5	65	400
6.0	68	425
6.7	72	450
7.0	75	475
7.5	79	500

Note: 2.8 g calcium and 2g phosphorus should be provided per kg of milk produced.

3. Growth

Table 3 : Nutrient requirements of growing cattle (growth rate at 550 g/day)

Live weight (Kg)	DM (Kg)	DCP (g)	TDN (Kg)	CA (g)	P (g)
70	2.10	259	1.39	8	5
80	2.33	282	1.53	9	6
100	2.78	328	1.80	12	9
120	3.23	373	2.07	15	11
140	3.67	419	2.34	17	12
150	3.90	442	2.47	20	13
160	4.12	465	2.61	20	13
180	4.57	510	2.88	20	13
200	5.02	556	3.14	20	13
220	5.47	601	3.41	22	15
240	5.97	647	3.68	25	17

Table 4 : Nutrient requirements of growing buffaloes (growth rate at 450 g/day)

Live weight (Kg)	DM (Kg)	TDN (Kg)
70	1.97	1.24
80	2.20	1.38
100	2.65	1.64
120	3.10	1.91
140	3.56	2.18
160	4.01	2.45
180	4.21	2.72
200	4.46	2.98
220	5.36	3.25

4. Nutrient Requirements for Working Cattle

Table 5 : Nutrients required for working animals per head per day*

Live wt.	Normal work #			Heavy work @		
(kg)	DCP(kg)	TDN(kg)	ME(Mcal)	DCP(kg)	TDN(kg)	ME(Mcal)
200	0.24	2.0	7.2	0.25	2.7	9.7
300	0.33	3.1	11.2	0.42	4.0	14.4
400	0.45	4.0	14.4	0.57	4.8	17.3
500	0.56	4.9	17.6	0.71	6.4	23.0
600	0.66	5.8	20.9	0.84	7.2	25.9

* Sen, K.C., S.N. Ray and S.K. Ranjhan (1995)
# working for 4hr of ploughing or 6hr of carting
@ Working for 6hr of ploughing or 8hr of carting

Exercises

1. Find out the nutrient requirements for maintenance of a cow weighing 350 kg body weight.
2. Calculate the nutrient requirements for a cow weighing 400kg yielding 20kg of milk with 4% butter fat per day.
3. Calculate the nutrient requirements for a buffalo weighing 450kg and yielding 15kg milk with 7% butter fat per day.
4. Find out the nutrient requirements for a pair of bullocks weighing about 400kg each and put to work for 4 hrs of ploughing per day.
5. Find out the nutrient requirements for a pair of bullocks each weighing about 500kg and working for 8 hrs of carting per day.

8. Formulation of Rations for Livestock

Ration formulation is very important aspect of animal production. The success of any animal production enterprise depends, to a large extent, on proper nutrition and feeding based on economical diets. The owner or animal production practioner should have a good knowledge of nutrition, feeding, physical and chemical characteristics of feedstuffs and feedstuff interactions and limitations, as well as the economics of production. Ration formulation is a task that matches nutrient requirements of the animal with combinations of various feed ingredients.

The information needed to balance a diet for any animal at any physiological stage includes the requirements of that animal and the chemical composition of the available feedstuffs. With this information in hand, the ration can be rapidly formulated.

- The first step is to know the nutrient requirements of a particular animal species for a particular physiological function, eg: growth, lactation. Nutrient requirements for animals are expressed in terms of energy, protein (amino acids), mineral elements and vitamins. In India, the ICAR (1985) recommendations are the most commonly used guidelines to work out the nutrient requirements of farm animals.
- The next step is to list all the available feedstuffs to be considered in the diet for a particular animal. Some feed stuffs may be limited

in use because they are indigestible or unpalatable. Therefore, both the feedstuff analysis and the maximum percentage allowances of a particular feedstuff for a diet can be considered while formulation of balanced rations.

- Generally diets are first formulated for one nutrient. Then the other nutrients are checked to see whether the feedstuffs used will meet the requirements or whether alternative feeds need to be included in the diet. One method is to balance protein first and then to check energy levels to see whether they are also met. Then the ration can be checked for other nutrients such as Ca and P.

Example

1. Formulate a ration for a cow weighing 400 kg, yielding 20kg fat corrected milk 4% (FCM) and in 1st lactation. Formulate a practical ration as per ICAR (1985) standards.

	Qty (kg)	DM (kg)	DCP (kg)	TDN (kg)
Requirements for maintenance	–	10-12	0.223	3.55
For production of 20kg FCM 55, 0.325	–	–	0.044	0.71
Total requirement (A)	–	10-12	1.367	10.76
Feeding Schedule				
NB21 green fodder (25, 1, 16)*	20	5.0	0.260	3.20
Paddy straw (90, 0, 40) *	4	3.6	–	1.60
Nutrients supplied through Roughage sources (B)	–	8.60	0.260	4.80
Balance of nutrients to be Supplemented from concentrate Mixture (A-B)	–	–	1.168	6.16

* DM, DCP and TDN on as fed basis

No. of kg of concentrate mixture required (containing 16% DCP) $= \dfrac{1168}{16} = 7.2$ kg

Exercises

1. Compute a ration for a Murrah buffalo weighing 500kg, yielding 15kg of milk with 7.5% milk fat when it is in 2nd lactation. Use all farm grown fodders like hybrid Napier, Lucerne, paddy straw and concentrate mixture.
2. Formulate a ration for a buffalo calf weighing 100 kg and growing at the rate of 450 g/day by using available feedstuffs.
3. Compute a ration for a crossbred cow weighing 400kg and at an advanced stage of gestation with the following locally available feedstuffs; Paddy straw, limited quantity of sun hemp hay, jowar green fodder and concentrate mixture.
4. Formulate a ration for feeding a breeding bull weighing 500kg. Choose the feeds that are available in your region.
5. Compute a ration for a pair of bullocks weighing about 400 kgs each when working for 4 hrs of ploughing per day. Choose the feedstuffs available locally.
6. Compute a ration for a pair of bullocks weighing about 500 kgs each when working for 8 hr of carting per day. Choose the feedstuffs available in your region.
7. In a semen bank there are 10 bulls with an average weight of 600 kg. The station can provide 15kg of Hybrid Napier (green) and 8 kg of paddy straw to each animal. How much concentrate mixture each animal requires and what should be the percentage of DCP in the concentrate mixture.

9. Principles of Compounding and Mixing of Feeds

Feed ingredients generally are mixed together in a commercially available mixer in the proper ration to provide a nutritionally balanced blend. There are two broad types of mixers, batch and continuous flow.

Batch Mixers

There are two configurations of batch mixers which may be either vertical or horizontal. Vertical mixers are cylindrical shape with an inverted cone-shaped bottom and one or more vertical augers to move feed from the bottom to the top of the mixer. A uniform mixture of all ingredients is attained within 15 to 20 minutes. Horizontal mixers are U shaped with paddles or ribbon like blades attached to a horizontal revolving shaft. Horizontal mixers can accommodate mixtures of chopped hay, silage, grains, molasses and other ingredients differing in particle size, density and texture. Mixing time is less than that of batch mixtures (2-15 minutes). To produce a desired quantity of mixed feed, individual feed ingredients are separately added to the mixer in the predetermined amounts. For most efficient mixing, usually the ingredients used in the largest amount is added first followed by ingredients that make up a smaller proportion of the final diet in descending order of their contribution to the total diet. When added fat is a constituent of the mixture, all other constituents are mixed first and the fat is added last and the final diet is mixed again.

Continuous Flow Mixers

Continuous flow mixers rely on continuously blending of all the ingredients in the proper ratios to produce the desired diet. Feed ingredients are simultaneously augured and blended. Proper calibration is essential to ensure that the desired composition is produced. The system once calibrated, can be operated automatically without direct intervention. There is less flexibility in changing feed ingredients than with batch mixers and most bulky ingredients such as hay and silage cannot be used. Continuous flow mixing systems with automated ingredients weighing and dispensing into the mixer are most commonly used in large poultry production units in which diet composition and ingredients are relatively constant.

Diet Formulation

Feed represents a large percentage of the cost of production; therefore, diet formulation is a very important aspect of animal production. Diet formulation allows the nutritionist to develop a diet that can be eaten and utilized by the animal to provide an economically feasible level of production. For diets that contain few feed stuffs and in which the nutritional parameters, such as energy and protein, are fixed, the techniques required are quite simple. In such cases use of algebric or Pearson's square method can be utilized. When diets become more complex and the optimum level of production is to be determined, mathematical programming is the technique of choice.

There are 3 ways of formulating the concentrate mixtures

1. Trial and Error Method
2. Pearson's Square Method
3. Algebric Method

1. *Trial and Error method :* In this method, one can formulate the concentrate mixtures through several trial and error methods of calculations. Unless one is very experienced it will take lot of time to formulate concentrate mixtures by this method.

2. *Pearson's square method:* In this method, desired percentage of protein is placed in the center of the square (16% in this example). The percentage of protein in the two feeds is to be put at the left hand corners of the square (8% CP for maize grain and 42% for GN cake). Then subtract (diagonally) the lesser percentage from the higher percentage values and place the answers in the right hand corners of the square (16-8 = 8 is placed at the top right and 42-16 = 26 is placed at the right bottom right).

Example : Prepare a concentrate mixture of DCP 16% and TDN 70% using maize grain, G.N. cake, rice bran and wheat bran.

Average DCP of high protein (GNC) = 42%

Average DCP of low protein (Rice bran + wheat bran + maize)

$$= \frac{6 + 10 + 8}{3} = \frac{24}{3} = 8$$

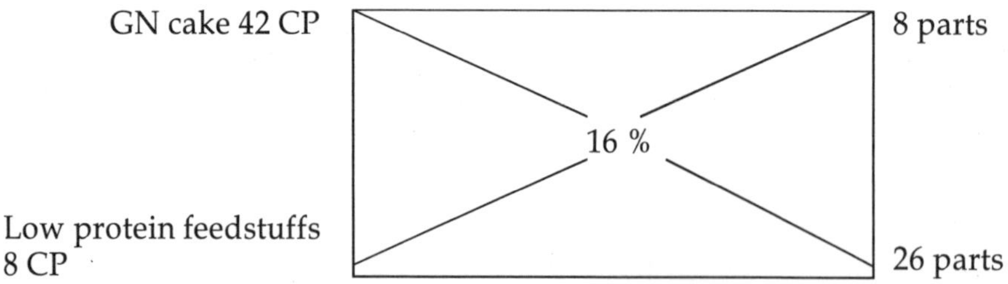

GN cake and low protein feeds have to be fixed in a proportion of 8:26 for 97 parts (2 parts mineral mixture and 1 part salt to be deducted from a total of 100 parts).

In 34 parts, high protein feed stuffs part is = 8

In 97 parts high protein feedstuffs parts =

$$\frac{8}{34} \times 97 = 23$$

The amount of low protein concentrate = 97 -23 = 74

They are to be mixed in a ratio of 6: 10:8

$$\text{Rice bran} = \frac{6}{24} \times 74 = 19$$

$$\text{Wheat bran} = \frac{10}{24} \times 74 = 31$$

$$\text{Maize} = \frac{8}{24} \times 74 = 24$$

Feed stuff	**Parts**	**Check % DCP supplied**	**% TDN supplied**
Ground nut cake	23	x 0.42 = 9.66	0.71 x 23 = 16.3
Rice bran	20	x 0.06 = 1.20	0.60 x 20 = 12.0
Wheat bran	30	x 0.10 = 3.00	0.68 x 30 = 20.4
Maize	24	x 0.08 = 1.92	0.80 x 24 = 19.2
Mineral mixture	2	--	--
Salt	1	--	--
Total	100	15.78	67.9

Therefore, depending on these calculations, above mentioned parts of feedstuffs can be mixed to formulate a concentrate mixture containing 16% DCP and 70% TDN.

Exercise

1. Prepare a concentrate mixture with a DCP of 16% and TDN 70% using maize grain, jowar grain, soybean cake and rice bran.
2. Formulate a concentrate mixture with 18% DCP and 65% TDN using the following ingredients. Sunflower cake, groundnut cake, jowar grain, rice bran and maize grain.

3. Prepare a concentrate mixture with a DCP of 13% and TDN 65% using locally available feed ingredients for inclusion in the maintenance type of ration for cattle.

4. *Algebric Method :* This method is to solve simple diet formulation using simultaneous equations. The amounts of the ingredients are unknown.

   Eg. : Prepare a concentrate mixture with a DCP of 16% and TDN 70% using the following ingredients. G.N. cake, rice bran, wheat bran, maize grain, 2 parts of mineral mixture and 1 part salt.

Let X is the amount of low protein concentrate in 100 kg feed.

Y is the amount of high protein concentrate in 100 kg feed.

Out of which 2 parts of mineral mixture and 1 part salt are already fixed to be included in the finally prepared concentrate mixture.

$X + Y = 100$ ------ (1)

Average DCP content of low protein content is 8%

Then number of kgs of protein in 100 kg feed = 0.08X

Average DCP content of high protein content is 42%

Then number of kgs of protein in 100 kg feed = 0.42Y

$0.08 X + 0.42 Y = 16$ --------- (2)

Multiply (2) by 100

$8X + 42 Y = 1600$ ---------- (3)

Multiply (1) by 8

$8 X + 8 Y = 800$ -------- (4)

Subtract the equation (4) from (3)

$8X + 42 Y = 1600$

$8 X + 8 Y = 800$

$34 Y = 800$

$$Y = \frac{800}{34} = 23.53 \text{ or } 23 \text{ parts}$$

• X = 74 parts {substitute Y value in equation (1)}

The amount of low protein concentrate = 97-23 = 74

They are to be mixed in the ratio of 6: 10: 8

$$\text{Rice bran} = \frac{6}{24} \times 74 = 19$$

$$\text{Wheat bran} = \frac{10}{24} \times 74 = 31$$

$$\text{Maize} = \frac{8}{24} \times 74 = 25$$

Feed stuff	Parts	% DCP supplied Check	% TDN supplied
Ground nut cake	23	x 0.42 = 9.66	0.71 x 23 = 16.3
Rice bran	20	x 0.06 = 1.20	0.60 x 20 = 12.0
Wheat bran	30	x 0.10 = 3.00	0.68 x 30 = 20.4
Maize	24	x 0.08 = 1.92	0.80 x 24 = 19.2
Mineral mixture	2	--	--
Salt	1	--	--
Total	100	15.78	67.9

Based on the above calculations, required quantities of feed stuffs can be mixed thoroughly to formulate a concentrate mixture containing 16% DCP and 70% TDN.

Formulating Premixes and Supplements

Premixes are used in animal nutrition in small amounts to supply vitamins, mineral elements, protein, amino acids, antibiotics and other feed additives. As with diet, premixes can be formulated using any of the techniques described earlier. When premixes need only a few ingredients the Pearson's square or simultaneous equations method can be used. The use of mathematical programming in premix formulation allows for the selection of the lowest cost raw materials within specified ranges.

BIS and CLFMA Standards for Compound Cattle Feeds

Two categories of cattle feeds viz., Type 1 and Type II have been prescribed by the Bureau of Indian Standards (BIS-formerly ISI).

	Type I **IS; 2052, 1979** **Re-affirmed 1990**	**Type II** **IS; 2052, 1979** **Re-affirmed 1990**
Moisture, % max	11	11
Crude Protein, % min*	22	20
Crude fat, % min	3	2.5
Crude fibre, % max	7	12
Acid insoluble ash, % max	3	4

• Except moisture, all on dry matter basis

The nutritional parameters prescribed for the Type I and Type II by BIS are such that they do not allow the incorporation of the abundantly available Non-conventional Feed stuffs (NCFS) and Agro-industrial Byproducts (AIBP) in this country. Introduction of one more type of cattle feed will facilitate usage of non-conventional feed ingredients, which will go a long way in bringing the existing gap between availability and demand for compound feeds.

The members of Compound Livestock Feed Manufacturers Association (CLFMA) have their own standards so that the feed manufacturers can produce quality compound feed to meet international standards. After careful consideration of views of several Animal Nutritionists, CLFMA standards are also now available for the formulation of cattle feed in India.

Based on body weight and milk yield dairy cattle were classified in to 3 categories:

Cows

i. Low yielder (Body wt. 350 kg, giving milk of 1-7 kg with average yield of 4 kg/day)

ii. Medium yielder (Body wt. 400 kg, giving milk of 8-14 kg with average yield of 11 kg/day)

iii. High yielder (Body wt. 450 kg, giving milk of 15-21 kg with average yield of 17 kg/day)

Buffaloes

i. Low yielder (Body wt. 400 kg, giving average milk yield of 4 kg/day)

ii. Medium yielder (Body wt. 450 kg, giving average milk yield of 8 kg/day)

iii. High yielder (Body wt. 500 kg, average milk yield of 12 kg/day)

The average feeding practices have been taken into consideration to arrive at the requirements through compound cattle feed. The production requirements of dairy cattle and buffaloes are based on the percentage of milk fat. Introduction of a third variety of cattle feed with 16% crude protein has been considered adequate for the traditional farmers who maintain 2 or 3 dairy cattle or buffaloes.

Item	Dairy special	Type I	Type II	Type III
Moisture, (% max)	10	11	11	11
Crude Protein, (% min)	22	20	18	16
On DM basis				
Undegraded Protein (% min)	8	--	--	--
Crude fat, (% min)	3	2.5	2.0	2.0
Crude fibre, (% max)	7	7	12	14
Acid insoluble ash, (% max)	3.5	4	4.5	5.0
Recommended levels of feeding (kg/day) Cow / Buffaloes				
Low Yielder :				
Maintenance				1.0(1.0)
Production (Kg/kg milk)				0.5(0.5)
Medium Yielder :				
Maintenance			1.0(1.5)	
Production (Kg/kg milk)			0.5(0.5)	
High Yielder :				
Maintenance		1.0(1.5)		
Production (Kg/kg milk)		0.4(0.5)		

* Values in the parenthesis are for buffaloes.

** For cows yielding 20kg milk; for maintenance 1.5 kg and for every kg milk yielded 0.4 kg of Dairy special.

*** 50g extra CP per kg milk for every additional 0.5 % milk fat over 4%.

While Type I and Type II cattle feeds meet the nutritional needs of medium and high yielding cows, the need for a cost effective Type III feed for the low milk yielder is clearly established. The feed manufacturers can compound the Type III feed using several of the unconventional feed resources and AIBPs in a balanced way using the available scientific knowledge about the incriminating factors present in some of the NCFRs and AIBPs. Type III cattle feed can be fed to work cattle doing normal and heavy works.

Exercise

1. Formulate dairy special concentrate mixture having 18% DCP with the following restrictions by using locally available feed ingredients and to contain:

Mineral Mixture	–	2 %
Salt	–	1%

2. Formulate a concentrate mixture required for adult cattle to have 18% CP with the following locally available feed ingredients:

   Groundnut cake, Gingily oil cake, Maize grain, Jowar grain, Deoiled Rice bran with a need to include Urea (2%); Molasses (8%); Salt (1%); Mineral mixture (2%) at levels shown in parenthesis.

3. Visit a local feed factory and note down your observations on important salient features of the factory, various feedstuffs that are used for making compound feeds for various species of livestock and the routine operations in the factory.

10. Least Cost Ration Formulation

Since feed cost represents 50 to 80 % of the cost of production, diet formulation using least-cost techniques is extensively used to minimize the cost of the diet. First, the nutrients and the minimum or maximum amounts of each nutrient are entered. For each feed to be considered, the current price, nutrient composition, and percentage limitations are also taken into account. Using this information, the computer picks the mixture of feed ingredients which will satisfy the nutrient levels specified. Once the appropriate equations are set up properly, any linear programming software package can be used to solve the diet formulation problems. After a solution is obtained, it is up to the nutritionist to evaluate and verify that the solution conforms to the nutritional knowledge i.e. the diet is palatable to the animal, does not contain nutrients and other substances in toxic amounts, and that the effects of other factors related to the specific nutrient requirements and feed consumption patterns of the animals for which the diet has been formulated will be optimal.

One limitation of conventional linear programming is nutrient density within the mix. If nutrient levels specified for mix are low, the computer will add a low cost ingredient as a filter material to reduce the cost of the ration. While the ration price may be very low, the filter may be of no benefit to the cattle. In this manner, a least cost mixture may not produce least cost production in cattle. When it is intended to formulate for maximum profit, other factors such as price of milk, and certain production costs are to be considered.

Exercise

1. Carefully follow the demonstration of formulation of least cost ration by using linear programming software in a personal computer for dairy cattle and note your observations in the record book.

11. Calculation of Nutrient Requirements of Sheep

Indian breeds of sheep have much lower growth rates and adult weights in comparison to many exotic sheep breeds. Besides the genetic make-up inadequate nutrition is the major factor responsible for such low growth and body weights. There is considerable scope for improvement of Indian Sheep by feeding them according to their needs as per the 'Nutrient Requirement' of ICAR (1985).

Growth

Nutrients to be provided for a lamb growing at three different rates at a particular weight after weaning (ranging from 10 to 25 kg) are presented in Table 1.

Table 1 : Nutrient requirements for growing lambs.

Live weight (kg)	Rate of gain (g/d)	DM (g)	DCP (g)	TDN(g)	Ca(g)	P(g)
10	50	400	35	220	2.0	1.5
	100	450	45	280	2.5	1.5
	150	500	55	340	3.0	2.0
15	50	500	45	300	2.8	2.0
	100	600	55	360	3.5	2.5
	150	700	65	450	4.5	3.0
20	50	700	50	400	3.5	2.5
	100	800	70	520	4.5	3.0
	150	1000	80	640	5.5	3.6
25	50	800	65	500	4.0	3.0
	100	1100	85	700	5.0	3.5
	150	1200	100	800	6.0	4.0

A pre-ruminant lamb is to be fed creep-feed of high quality according to appetite from its age of 2 weeks to weaning at 90 days to promote growth during early age through optimum rumen development. Examples of formulating of creep-feed for pre weaned lambs are given in Table 2.

Table 2 : Creep mixture for young lambs

S.No.	Ingredients	% (air dry basis)
1.	Maize /barley	50.0
	Groundnut cake /Linseed cake	20.0
	Wheat bran / Rice bran	22.0
	Fish meal	5.0
	Mineral Mixture	2.5
	Common salt	0.5
	Total	100
2.	Maize /barley	30.0
	Gram	10.0
	Groundnut cake /Linseed cake	32.0
	Wheat bran / Rice bran	25.0
	Mineral Mixture	2.5
	Common salt	0.5
	Total	100
3.	Maize /barley	30.0
	Bajra	10.0
	Groundnut cake	22.0
	Groundnut lucerne / berseem	10.0
	Fish meal	5.0
	Wheat bran / Rice bran	20.0
	Mineral Mixture	2.5
	Common salt	0.5
	Total	100

> **Note**
> 1. Grains should be broken or coarsely ground.
> 2. Oil meals should be coarsely ground.
> 3. Vitamin A and D supplements to be added at the rate of 30g/q of feed or green fodder should be allowed free choice to lambs.

Maintenance

The maintenance requirement values have been arrived at from several digestibility and balance trials in which animals maintained their body weight. The DCP requirement has been found to vary from 0.9 to 1.2 g per kg live weight, while the DCP to TDN ratios varied from 1: 10-12. DCP and TDN requirements are 2.73 g/kg $W^{0.75}$ and 273 g/kg $W^{0.75}$ respectively.

Table 3 : Nutrient requirements for maintenance of adult sheep

Live weight (kg)	DM (g)	DCP (g)	TDN (g)	Ca (g)	P (g)
20	500	25	240	3.0	1.8
25	625	30	300	3.2	2.0
30	750	32	360	3.5	2.5
35	875	35	420	4.0	2.8
40	1000	40	480	4.5	3.0
45	1125	42	500	5.0	3.5
50	1250	45	540	5.5	3.8
55	1375	50	600	6.0	4.0
60	1450	54	640	6.5	4.5

Reproduction

Though no specific information is available on the nutrient requirements of sheep for spermatogenesis and breeding efficacy, there is evidence that deficiency of one or more nutrients in the diet may affect these traits. It is therefore, recommended that breeding rams should be provided 50 per cent more nutrients than what is recommended for their maintenance. Similarly, pregnant ewes during the last 6 weeks of gestation should be provided with 50 per cent more nutrients than the maintenance needs. Sheep also require extra nutrients to bring them to flush and, therefore, should be provided 25% more nutrients above the maintenance needs.

Wool Production

No specific recommendations are given for wool production as it has been observed that if sheep are provided all the essential nutrients required for maintenance, growth and production, they should also meet the requirements for wool production.

Exercise

1. Suggest a good creep feed for pre-ruminant lamb using the feedstuffs available locally.
2. Calculate the nutrient requirements of a lamb weighing 15 kg and growing at the rate of 100g per day and suggest an ideal feeding schedule.
3. Formulate a ration for sheep on an average weighing 30kg and in advance stage of pregnancy which is maintained under intensive system.
4. Formulate a ration for a breeding ram weighing 50kg which is maintained under intensive system.

12. Nutrient Requirements of Goats

Goats can well utilize the coarse fibrous feeds (including FCR) and are higher in efficiency of cellulose digestion and feed utilization for milk production. The DM intake of goats is higher in comparison to large farm animals. Meat goats have a DM intake of 3 % of their body weight whereas dairy goats have a higher DM intake. Goats are known for their round the clock eating habit. Due to smaller GI tract capacity the daily feed allowance needs to be offered 3 or 4 times. On an average a goat requires 3.5kg of green fodder per day.

Growth

The requirements for growth have been prescribed for kids weighing 10 to 30 kg and growing at the rate of 50g/day and are presented in Table 4.

Table 4 : Nutrient Requirements of kids

Live weight (kg)	DM (g)	DCP (g)	TDN (g)	Ca (g)	P (g)
10	425	25	275	1.8	1.1
15	600	30	350	2.0	1.2
20	700	35	400	2.4	1.4
25	800	40	450	2.6	1.5
30	950	45	500	2.9	1.6

Maintenance

The dry matter requirements of goats have been found to be $76g/W^{0.75}$, and that of DCP, TDN, Ca and P as 2.5g, 30.0g, 0.145 g and 0.072 g per kg metabolic size. These figures are calculated for animals weighing 30 to 70 kg and are listed in Table 5.

Live weight (kg)	DM (g)	DCP (g)	TDN (g)	Ca (g)	P (g)
30	900	35	400	2.0	1.0
40	1200	40	500	2.5	1.2
50	1400	45	560	2.7	1.4
60	1600	55	670	3.0	1.6
70	1800	60	760	3.5	1.8

Reproduction

For meeting the demands of the growing foetus in advanced pregnancy and spermatogenesis in males during breeding season, an extra allowance of 55g DCP, 400g TDN, 2g calcium and 1.4g phosphorus should be provided over and above maintenance requirement.

Milk Production

For each litre of milk produced, the dairy goat should be provided with 70g DCP, 350g TDN, 3.0g calcium and 2.1g phosphorus. Goats in milk have very high requirements for sodium chloride.

Exercise

1. Suggest an ideal creep feed for pre-ruminant kid using feed stuffs available locally.

2. Calculate the nutritional requirements for a kid weighing 15 kg body weight and growing at the rate of 50g per day and suggest ideal feeding schedule.
3. Formulate a ration for a goat weighing 30kg and at advanced stage of last quarter of pregnancy which is maintained under intensive system of rearing.
4. Compute a ration for a breeding buck weighing 50kg which is maintained under intensive system.
5. Formulate a ration for a crossbred goat weighing 30kg and yielding 1.5 litres of milk per day which is maintained under intensive system of rearing.
6. Formulate a ration for a Saanen goat weighing 40kg and yielding 2 litres of milk per day under intensive system of rearing.

13. Feeding of Livestock During Scarcity and Disastrous Conditions

Scarcity of animal feed caused by floods is of a temporary nature. During drought, severe shortage of animal feed, especially of roughages is encountered. This has an adverse effect on feed supply position, as dried roughages alone account for over 50 to 80 per cent of supply of dry matter and TDN required for farm animals. Failure of monsoon is the main factor responsible for droughts in this country. The use of byproducts of commercial crops such as oil seed crops, grams etc., as well as of drought resistant vegetation, in combination with NPN source of nitrogen and of molasses as source of energy, can be used for meeting the immediate nutritional requirements under conditions of scarcity. Certain varieties of tree leaves and extractions of inedible oil seeds can be used to meet the essential requirements of intact protein in ruminants. This strategy was successful during droughts in many parts of India. Cattle camps were set up around sugarcane factories located in the drought affected zones. Large-scale feeding of bagasse and molasses in combination with urea and mineral supplements was carried out without any detrimental effects. Some of the daily ration schedules that have been successfully tried are given below:

i. Ration for adult non-producing cattle (50-300kg body weight) fed twice a week.

Bagasse	-	2.0kg
Molasses	-	0.4kg
Sugarcane tops*	-	8.0kg (chopped)
Urea	-	22g
Common salt	-	30g
Sterilized bone meal		
or mineral mixture	-	20g
Trace element mix	-	0.5g

* In case of non-availability of tops, the ration should be fortified with 5,000 IU of Vitamin A and replacing the tops with forest grass hay or any other available crop residues in surrounding areas of drought affected zones.

ii. Ration for a growing animal (50-150 kg body weight) fed twice a week:

Bagasse	-	2.0kg
Molasses	-	0.8kg
Sugarcane tops*	-	3.0kg (chopped)
Urea	-	40g
Common salt	-	20g
Sterilized none meal		
Or mineral mixture	-	20g
Trace element mix	-	0.5g

The feed can be prepared by as simple method and using available farm equipment. Bales of bagasse are spread in layers of 15-20 cm thickness; urea and molasses are mixed in empty drums. Approximately, 23 per cent of water is added to reduce dustiness of material. The urea-molasses mixture is sprayed over the bagasse, which is turned frequently to ensure proper mixing and soaking. For spraying urea molasses mixture watering can is used. This process can be repeated using dry tree leaves, coarse forest grasses and such other materials available around the areas of short supply.

Other New Systems of Feeding

i. Liquid urea-molasses feeding
ii. Uro-mol Brick or Block or Lick.

i. *Liquid urea-molasses feeding* : Liquid molasses containing 2 to 3 per cent uniformly mixed with urea and fortified with minerals and vitamins is named as 'Liquid feed'. The principle of urea molasses liquid feed preparation is the homogenous mixing of urea in liquid molasses. A most practical and least toxic urea molasses liquid contains 2.5 parts urea, 2.5 parts fresh water, 2 parts mineral mixture, 1 part common salt and 92 per cent sugarcane molasses. The mixture is fortified with vitablend (vitamin A and D_3) at the rate of 25g per 100 kg of liquid feed. Urea is completely dissolved in water and poured gradually with a simultaneously mixing in the tank with a wooden spoon. Powdered salt and mineral mixture are sprinkled over the molasses while mixing to ensure uniform distribution of all the additives in the liquid molasses. Special attention is required for the uniform mixing of urea solution. Since in winter the viscosity of molasses increases, heating of the mixture for thorough mixing is required. Undiluted urea molasses liquid feed containing 65 per cent and more of dry matter can be safely stored for long.

Another combination of liquid urea molasses feed is as follows:

Molasses	-	83%
Urea + Phosphoric acid	-	4%
Salt	-	2%
Mineral Mixture	-	1%
Bypass protein	-	10%

Precautions to be Followed while Feeding Liquid Urea-Molasses Diets

- It is desired to supplement the liquid molasses diet to minimum quantity of roughages to maintain bulk.
- It is advisable to introduce liquid feed gradually in the ration in about 10 to 15 days.
- Free choice of fresh drinking water should be available for the animals on liquid diet.
- It is advantageous to provide free choice of mineral mixture/ mineral licks in a separate container.

ii. *Uro mol Brick feeding :* The National Dairy Development Board (NDDB) has started manufacturing urea molasses block. The production and utilization of urea molasses blocks assumes an increasingly important role in this drought prone sub-continent. When licked by the ruminant animals, the uro molasses blocks provide nitrogen to micro-organisms in the rumen and in turn improve digestion of the straw that is fed during disastrous conditions. These blocks are available commercially in packed condition so that they can be transported to the areas of necessity easily.

The composition of urea molasses mineral block is as follows:

Molasses	-	45%
Urea	-	15%
Mineral mixture	-	15%
Salt	-	8%
Calcite powder	-	4%
Bentonite	-	3%
Any soluble cake	-	10%

Molasses, urea, mineral mixture, salt and calcite powder are mixed with thorough stirring and boiled. Binder is added to the cake and the mixture is poured into the block molds and allows it to cool. Apart from cake, bran or some roughage source can also be incorporated to prepare complete feed blocks for feeding of ruminant livestock in drought prone areas. Nowdays dewormer included medicated blocks are commercially available for feeding to small ruminants.

Exercise

1. Identify the feed stuffs that are available in your area that can be best utilized for feeding livestock during disastrous conditions and give their nutritive value.
2. Give details of some commercially available urea-molasses blocks for feeding of livestock.

14. Computation of Rations for Swine

Nutritive requirements of pigs in the temperate regions of economically advanced countries are based on cereal grains like corn, wheat, barley or oats. Such knowledge on requirements of pigs in the tropical and subtropical zones is limited where cereal grain feeding is a luxury and ill afforded in practical swine feeding.

The quantitative nutrient requirements for swine during various stages of life cycle are summarized in Table 1 (ICAR, 1985). Swine tend to eat to meet energy needs; therefore concentrations of specific nutrients in the diet are generally expressed in units per kilogram of total dry diet consumed ad libitum. The energy component of the diet is the largest single constituent and is therefore of major concern in diet formulation.

Growing and Finishing Pigs

The intake by the growing and finishing pigs mainly depends on the energy content of the ration. Therefore, it is essential to standardize it for various categories of pigs. If the pigs are fed according to the requirements shown in the following table, they are expected to gain at a rate of 180, 335, 400, 490, 520 and 540 g, respectively during the body weight changes of 5-10, 10-20, 20-30, 30-40, 40-50 and 50-60 kg on feed intake levels of 0.25, 1.05, 1.30, 1.80, 2.10 and 2.40 kg per day. The ingredients can be replaced by other unconventional sources of protein and energy by keeping up the desired nutrient concentration in that particular type of diet.

Table 1 : Nutrient content of diets for growing and finishing pigs.

	Live weight class (kg)					
Feed unit	5-10	10-20	20-30	30-40	40-50	50-60
CP (%)	18	16	16	16	14	14
TDN (%)	70	68	68	68	68	68
DE(Kcal/kg)	3,100	3,000	3,000	3,000	3,000	3,000
Ca (%)	0.8	0.6	0.6	0.6	0.5	0.5
P (%)	0.6	0.5	0.5	0.5	0.4	0.4
Common salt (%)	0.6	0.5	0.5	0.5	0.5	0.5

Breeding Pigs

Breeding pigs can successfully be maintained on diets prescribed for the finishing pigs in the body weight range of 40 kg and above when fed at the rate of 2 per cent of body weight; lactating sows should be fed at the rate of 3 per cent of body weight, and gestating sows and gilts should be fed at the ratio of 1.5 and 2 per cent of their body weight.

Ration Formulae for Different Kinds of Pigs

Some examples of ration formulae from the feed ingredients available in different parts of India have been given for the feeding of pigs at different stages of growth and production.

1. Prestarter Ration

The rations of high protein and energy content offered to piglets from second week of life upto weaning (6-8 weeks) are pre-starter rations. Adequate amount of good quality protein of animal origin is very important for balancing the rations (Table 2). Suckling piglets under intensive management are prone to anaemia in early life due to deficient supply of iron in dam's milk, thus iron preparations are injected during the first and third weeks of life. The pre-starter feed is offered in separated feeders for piglets protected from the reach of the dam and other pigs. Piglets learn to eat in 2-3 days period. Their voluntary intakes depend on the milk production of dam and the palatability of the rations.

Table 2 : Composition of pre-starter rations

	Percent content in ration		
Constituents	**1**	**2**	**3**
Maize flour	40	30	20
Barley/Jowar/ragi flour	--	10	20
Skim milk powder	10	10	10
Fish meal	6	6	6
Groundnut cake	30	30	30
Molasses	10	10	10
Brewer's yeast	2	2	2
Mineral mixture	1.5	1.5	1.5
Common salt	0.5	0.5	0.5
Vitamin pre-mix	10g	10g	10g

2. Starter Rations

The ration offered to piglets from the day of weaning to about 20kg body weight or 3-4 weeks period is almost similar to the prestarter ration except that the skim milk powder is replaced by either fish meal or meat meal. If skim milk powder is available at cheaper rate, prestarter rations can also be used.

3. Grower Rations

The rations fed to growing piglets from 20 kg body weight to 35 kg body weight of exotic pigs are called grower rations. These rations contain lower level of animal proteins and energy and during this phase, it is possible to replace larger proportion of cereal grains with grain processing by-products and several agro-industrial waste materials. Few examples of cereal based and cereal replaced grower rations are given in Table 3.

Table 3 : Composition of grower rations

	Percent of ration						
Ingredient	**1**	**2**	**3**	**4**	**5**	**6**	**7**
Crushed maize	60	40	20	30	--	--	10
Jowar	--	10	20	--	--	--	10
Minor millets	--	10	--	--	--	--	10
Groundnut cake	20	20	20	15	15	15	15
Sunflower cake	--	--	--	10	--	--	10
Wheat bran	10	10	15	20	75	50	
Deoiled rice bran	--	--	15	15	--	25	
Fish meal/Meat meal	8	8	8	8	8	8	8
Mineral mixture	1.5	1.5	1.5	1.5	1.5	1.5	1.5
Common salt	0.5	0.5	0.5	0.5	0.5	0.5	0.5
Vit. supplement	10g	10g	10g	10g	10g	10g	10g

Replacement of cereal grains results in proportionate fall in the energy content of the ration resulting in lowered growth response in pigs. Some energy requirement is compensated through increased intake of low energy feeds.

4. Finisher Rations

The rations fed to fattening pigs from 35 kg live weight to slaughter at 70-90 kg body weight are known as finisher ration. Slaughter of exotic pigs at 70kg body weight has been suggested in the Indian conditions due to economic reasons. Finisher rations contain lower level of protein and fish meal.

Table 4 : Comparison of finisher rations

Ingredient	Percent of ration					
	1	2	3	4	5	6
Crushed maize	50	30	20	--	--	10
Sorghum/Bajra/Ragi	--	20	20	--	--	15
Wheat bran	18	18	28	65	40	13
Rice bran	--	--	--	--	25	25
Groundnut cake	15	15	15	10	10	10
Sunflower cake	--	--	--	--	--	10
Fish meal	5	5	5	5	5	5
Molasses	10	10	10	18	18	10
Mineral mixture	1.5	1.5	1.5	1.5	1.5	1.5
Salt	0.5	0.5	0.5	0.5	0.5	0.5
Rovimix	10g	10g	10g	10g	10g	10g

The finisher ration can be modified according to the availability of feedstuffs and their feeding values for the pigs. Significant increase in voluntary feed intake occurs on the feeding of low energy rations containing palatable feeds.

5. Feeding of Pregnant and Lactating Sows

Restricted feeding to support about 20-25 kg gain in body weight of pregnant gilts and sows has been found to be optimum, and pregnant sows should be fed about 2 kg mixed feed daily supplying about 300g protein and 6600 Kcal DE. The lactating sows are fed a similar ration at the rate of 3-4 kg daily depending upon the body size. In addition to this piglet allowance is offered at the rate of 200g per piglet daily during 6-8 weeks of lactation period. The ration is drastically reduced to 2kg daily during the last week of weaning period for the drying of the sows.

6. Feeding of Breeding Boars

Breeding boars can maintain good vigor on the feeding of the grower ration at the rate of 3.0 kg before 18 months of age, and 2.5 kg to mature boars. The rations should contain adequate minerals and vitamins. Feeding of 4-5 kg succulent green fodders keep the breeding stock active and vigorous.

Exercise

1. Formulate a creep feed for piglets using locally available feedstuffs.
2. Compute a grower ration for piglets using unconventional feed ingredients.
3. Formulate a finisher ration for pigs using unconventional feed ingredients.

Table 5 : Nutritive values of feed stuffs that are commonly fed to pigs.

Feed Ingredients	Crude Protein (%)	Digestible Energy (Kcal/kg)
Maize grain	8.5	3530
Jowar grain	8.9	3280
Bajra grain	11.7	2700
Barley	11.5	3120
Ragi grain	11.0	3100
Wheat broken	16.5	3080
Deoiled Rice bran	14.0	3135
Rice polish	11.9	3713
Wheat bran	15.5	2370
Tamarind seed waste	15.4	2600
Cotton seed meal	41.7	2670
Groundnut cake, solvent ext.	49.0	3140
Niger seed cake	33.2	2600
Safflower meal, solvent ext.	22.9	2960
Sunflower cake, Solvent ext.	26.2	3100
Sesame (Til cake, Solvent ext.	45.0	3385
Soybean cake	44.0	3490
Skimmed milk powder	33.3	3845
Fish meal	43.1	2200
Meat meal	55.6	2805
Blood meal, dried	86.0	2980
Lucerne meal	17.4	1880
Molasses	4.4	2210
Yeast brewer's dried	43.8	3295

15. Ration Formulation for Poultry

Poultry has assumed great significance due to its ability to provide high quality food for human consumption within its, short life-span. The performance of the germplasm imported from temperate regions is poor under tropical environment because of low consumption and poor feed quality. Therefore, it is essential to prescribe the nutritional requirements of poultry under the environment they are being raised.

Poultry need energy for growth, maintenance of body tissue, production of eggs, regulation of body temperature and activity. The requirements of poultry for protein, Ca, P, Linoleic acid and other nutrients are determined mainly by the rate of accumulation of nutrients in body tissues or eggs and by the obligatory losses of nutrients that are associated with maintenance. Consequently, for any given age, body weight, and productive function (eg. growth, egg production), the daily intake requirements for these nutrients are relatively constant.

Feed intake is determined primarily by the bird's need to obtain energy. If the metabolizable energy (ME) requirement of the bird and the ME content of its feed are known, feed intake can be predicted. A laying hen, for example, requiring 300 Kcal of ME per day will consume (300 kcal ÷ 3000 kcal/kg) 0.100 kg (100g) of a diet containing 3000 kcal/kg of ME. If the daily amino acid requirements (i.e., milligrams of amino acids required per day) are known, then the amounts needed per kilogram of diet can be calculated. Let's assume that we wish to formulate for lysine, the requirement of which is 750 mg/hen/day. In our example of a hen consuming 100 g/day of a diet containing 3000 Kcal of metabolizable energy/kg, this represents a dietary lysine concentration of (0.75 g lysine/0.1 kg of diet) 7.5g lysine/kg of diet. If the energy density of the diet is altered, the amount of feed consumed each day will change, and this will affect the concentration of lysine that must be provided in the diet. If the energy density of the diet is decreased from 3000 to 2700 Kcal/kg, for example, feed intake is expected to increase to (300 Kcal ÷ 2700 Kcal/kg=) 0.111 kg (111g), and the lysine content of the diet can be safely decreased to (0.75g ÷ 0.111 kg) 6.76 g/kg of diet. Conversely, if the metabolizable energy density of the diet increases to 3300 Kcal/kg, then feed intake is expected to decrease to (300 kcal ÷ 3300 kcal) 0.09/kg/day, the concentration of amino acids in the diet must be increased to (0.750g

÷ 0.091 kg) 8.24 g/kg of diet in order to provide for the same daily intake as achieved when the hen receives the diet containing 3000 kcal/kg of metabolizable energy. An example of the methionine and lysine requirements of the layer at different dietary energy levels is provided in the following table.

Table 1 : Dietary methionine and lysine levels for layers in diets of varying energy densities.

ME (Kcal.kg)	Feed intake (g/hen/day)	Methionine (%)	Lysine (%)
2600	115	0.30	0.65
2800	107	0.33	0.70
3000	100	0.35	0.75
3200	94	0.37	0.80

Similar interrelationships between energy and nutrient levels may be assured to apply for protein and other macronutrients such as Ca, P and linoleic acid. They also could be applied to vitamins and trace elements. However these nutrients are added to the diet in small quantities and are relatively inexpensive. They routinely are provided in the diet with sufficient margin of safety so that requirements are satisfied inspite of variations in feed intake. Adjustments with regard to energy requirements usually are not made. The requirements based on the optimum performance of the birds have been given by ICAR (1985) for protein, metabolizable energy and other important nutrients and are presented in Table 2.

Table 2 : Nutrient content of diets for broiler, layer and breeder poultry

Nutrient	Broiler Starter	Finisher	Layer Starter	Grower	Layer	Breeder
Metabolizable energy(Kcal.kg)	2800	2900	2700	2500	2850	2850
CP (%)	23	20	20	16	18	18
Linoleic acid (%)	1	1	1	1	1	1
lysine	1.2	1.0	0.9	0.6	0.65	0.65
Minerals						
Calcium (%)	1.0	1.0	1.0	1.0	3.5	3.5
Available P (%)	0.45	0.45	0.45	0.45	0.5	0.5
Sodium (%)	0.2	0.2	0.2	0.2	0.2	0.2
Potassium (%)	0.4	0.4	0.4	0.4	0.4	0.4
Manganese (%)	70	70	70	70	70	70
Iron (mg/kg)	80	80	80	80	80	80
Copper (mg/kg)	8	8	8	8	8	8
Zinc (mg/kg)	40	40	40	40	60	60
Iodine (mg/kg)	0.6	0.6	0.6	0.6	0.5	0.5

contd...

Table 2 contd...

Vitamin (per kg)

Vitamin A (IU)	2000	2000	3000	3000	5000	5000
Vitamin D3 (IU)	600	600	600	600	1000	1000
Vitamin E (mg)	20	20	20	10	10	20
Vitamin K (mg)	2	2	2	2	2	2
Thiamin (mg)	4	4	4	2	2	2
Riboflavin (mg)	5	5	5	5	5	8
Pantothenic acid (mg)	15	15	15	15	15	15
Nicotinic acid (mg)	40	40	40	20	20	20
Pyridoxine (mg)	6	6	6	6	6	9
Biotin (mg)	0.2	0.2	0.1	0.1	0.1	0.15
Folic acid (mg)	2	2	1	1	0.5	0.5
Vitamin B12 (mg)	0.02	0.02	0.009	0.005	0.003	0.005
Choline (mg)	1500	1500	1300	--	--	--

Broiler

From economic point of view, the broiler is required to be slaughtered at an early age. Thus, the growth rate attained by these birds is higher than other categories of poultry, and these birds are to be provided nutrients in a concentrated form. While the requirements of protein and energy are higher, the ratio of these nutrients is lower in case of broilers.

Layer

The usual goal in the feeding of laying hens is to obtain the maximum production of eggs of large size and good internal and external quality with a least cost feed formulation. The laying birds are to be provided nutrients in such a way that they grow at a satisfactory rate, but do not fatten. Thus, the nutrient content of rations is different from other categories of the birds. Layer rations are to be provided in three different types, the specifications of which are included in Table 2.

Breeder

If the layers constitute a breeding population, the goal is to obtain maximum production of eggs of an appropriate size for incubation with good potential for high yield of healthy and vigorous chicks. The nutritional requirements of breeder (Table 2) populations are not greatly different from those of commercial layers; generally they require higher levels of vitamins and minerals to ensure that the egg is sufficiently enriched in these nutrients to support optimal development and hatchability of the chick (NRC, 1994).

Dietary Amino Acid Levels for Poultry

The amino acid requirements of meat and egg type chicken at different age groups are shown in Table 3. These are expressed as per cent of the diet, and the values are based on levels of these amino acids in rations that are found to yield optimum responses under practical feeding conditions (ICAR, 1985).

Feed Formulation

1. The following information is required for feed formulation:

a. Requirement of nutrients for poultry and composition of a formula in terms of nutrients.

b. Feed analytical values- The average nutrient content of the feed ingredient based on the analyses done is available in the form of published tables called "Feed Composition Tables". The feed composition tables compiled for poultry by the Central Avian Research Institute, Izatnagar, are revised and are reproduced in Annexure-I.

c. Maximum level of inclusion of feed ingredients- Each ingredient has its maximum level of inclusion in the diet. Inclusion of an ingredient beyond the maximum level may induce imbalance of nutrients, difficulty in feed formulation, and may reduce the performance of the birds due to the presence of anti-nutritional factors beyond tolerance level. Cost is one of the factors imposing an upper limit on the inclusion of feed ingredients. Table 3 indicates the maximum level of inclusion of common feed ingredients in poultry diets.

Table 3 :

Amino acid	Broiler starter	Broiler finisher and chick starter	Grower	Layer and breeder
Arginine	1.33	1.16	0.93	0.79
Glycine and / or serine	1.10	0.96	0.77	
Histidine	0.44	0.38	0.30	0.31
Isoleucine	0.83	0.72	0.58	0.50
Leucine	1.54	1.34	1.07	1.21
Lysine	1.22	1.06	0.85	0.50
Methionine+Cystine	0.83	0.72	0.58	0.54
Methionine	0.44	0.38	0.31	0.29
Cystine	0.39	0.34	0.27	0.25
Phonylalanine+Tyrosine	1.45	1.26	1.00	0.99
Phenylalanine	0.78	0.68	0.54	0.54
Tyrosine	0.67	0.58	0.46	0.45
Threonine	0.78	0.68	0.54	0.41
Tryptophan	0.23	0.20	0.16	0.13
Valine	0.94	0.82	0.66	0.65
Protein	24.0	19.0	16.0	17.0

Availability and cost of feed ingredients - In computer feed formulation, many nutrients can be taken into consideration at a time to formulate least cost rations through linear programming by utilizing the data of various feedstuffs already entered (Table 4).

Table 4 : Maximum level of inclusion of feed ingredients in poultry diets

Ingredients	Level of inclusion* (%)
Energy sources	
Maize	60
Wheat	50
Barley	20-40
Oats	10-20
Sorghum (dark variety)	10-20
Sorghum (white variety)	25-40
Pearl millet	50
Rice	40
Triticale	40
Rice bran	10-20
Deoiled Rice bran	10-15
Wheat bran	10-15
Maize bran	10
Sal seed meal (deoiled)	3-5
Molasses	5-10
Animal and vegetable fat	10
Poultry manure meal	10
Tapioca Flour	10-20
Gram Chuni	10-15
Leaf-meals	
Alfalfa-leaf-meal	5
Groundnut leaf meal	5
Protein Sources	
Groundnut cake	40
Groundnut cake (deoiled)	20
Soybean meal	40
Sesame meal	20
Safflower meal	20

contd...

Table 4 contd...

Linseed meal	4
Linseed meal (water treated)	20
Coconut meal	5
Cottonseed meal (degossypolized)	5
Mustard meal	10
Maize gluten meal	15
Niger cake	10
Fish meal	10
Fish scrap	5
Meat Meal	10
Meat cum bone meal	5
Blood meal	3
Poultry by-product meal	5
Silk worm pupae meal	6
Feather meal	2
Skim meal (dried)	3
Fermentation by -products	
Distiller's Yeast (dried)	5
Yeast sludge (dried)	10
Mahua cake	8

* When the values are given as range, lower levels of range are for chicks and higher levels of range are for layers.

II. Steps in Feed Formulation

In practical diet formulation by calculations, the essential steps are as follows: The ingredients are added to the formula to provide the nutrients per kg or 100kg.

Steps I- The minor ingredients are fixed and or slack space may be left to include them later. These are added in limited quantities based on cost or to provide a specific nutrient or non-nutrient feed additive or to balance a nutrient usually deficient. Trace mineral, vitamins and feed additives can be fixed because the contribution of major feed ingredients for these nutrients is little. Slack space may be left for addition of common salt, calcium and phosphorus sources, supplemental amino acids and fats at a later stage to balance the diet.

Fixed minor ingredients – Trace minerals, Vitamins, Coccidiostats, Antibacterial agents

Slack space for – Fat, Phosphorus source, Calcium source, Salt, Amino acids and Mineral mixture.

Fixed minor ingredients and slack space- 5kg

Step 2 – The levels of animal protein sources are fixed. These are added at fixed levels because of cost. These provide the limiting amino acids (lysine, methionine and cystine) at higher levels than other feed ingredients. Step –3 Vegetable protein sources and energy sources (cereals and by-products) are added to provide the required amount of protein.

Animal protein sources	-	Fish meal (43% CP), 5kg Meat meal (56% CP), 5 kg
Cereal by-products	-	Rice polishing –10 kg Deoiled rice bran – 10kg

Till now 35kg of ingredients are added (including slack space) and these contributed 7.37 per cent protein. Still 65kg of ingredients are to be added and 9.63 (17-7.37) per cent protein is to be made up. Groundnut cake and sunflower cake as vegetable protein source and maize, jowar as energy source are considered. The required protein level can be calculated by algebric equation or by Pearson's square method as already discussed in earlier chapters.

Algebraic Equation

Total ingredients	:	65 kg
Protein required	:	9.63 %

Let X represents the protein source and Y represent energy source. The average protein content of protein source is 42.9 and that of energy sources is 9.75 per cent.

$X + Y = 65$ – – – – (i)

$0.429X + 0.975Y = 9.63$ – – – (ii)

Multiply equation (2) by 100

$42.9\ X + 9.75\ Y = 963$ – – – – (iii)

Multiply equation (1) by 9.75

$9.75\ X + 9.75\ Y = 633.75$ – – – – (iv)

Subtract equation (iv) from equation (iii)

$$42.90\,X + 9.75\,Y = 963.00$$

$$9.75\,X + 9.75\,Y = 633.75$$

$$31.9\,X = 329.75$$

$$X = \frac{329.75}{31.90} = 10.34 \text{ or say 10 parts}$$

$$Y = 65 - 10 = 55$$

$$\text{Part of groundnut cake} = \frac{48.6}{85.8} \times 10 = 6$$

$$\text{Part of sunflower cake} = \frac{37.2}{85.8} \times 10 = 4$$

Parts of energy supplements to be mixed

$$\text{Part of Maize} = \frac{9}{19} \times 55 = 26 \text{ parts}$$

$$\text{Part of Jowar} = \frac{10}{19} \times 55 = 29 \text{ parts}$$

Table 5 : The following is the composition of the diet with 17% protein for layers

Name of the feed ingredients	Parts /100kg	CP	ME (Kcal/kg)
Maize	26	0.09x26=2.34	33.09x26=860
Jowar	29	0.1x29=2.9	26.45x29=767
Rice polish	10	0.12x10=1.2	29.37x10=294
DORB	10	0.14x10=1.4	22.35x10=224
Groundnut cake	6	0.48x0.6=2.88	23.28x0.6=140
Sunflower cake	4	0.37x0.4=1.48	22.30x0.4=89
Fish meal	5	0.43x0.5=2.15	18.34x0.5=92
Meat meal	5	0.56x0.5=2.8	23.19x0.5=116
Mineral mixture	1.5	--	--
Salt	0.5	--	--
Lime stone (Shell grit)	3	--	--
Vitamin mixture	*	--	--
Total	100	17.15	2722

* As per the direction of the manufacture

Step 5 - The metabolize energy content of the diet has to be balanced. The shortfall of 128 kcal (2850-2722 kcal) can be met by supplementation of 16.6g of animal fat. The cost of supplemental fat may be so high as to prohibit its inclusion in poultry diets. In such circumstances one has to compromise with inclusion of cereals particularly maize as energy source to increase the ME content of the diet.

Step 6 - A check is made for the total of the ingredients and also for all the nutrients if desired. If the total of the ingredients is not 100, cereal or cereal by-products can be added to make it 100. The diet may be modified by incorporating higher levels of cereals to meet energy of higher levels of vegetable or animal protein to meet the protein or specific amino acid requirements, if necessary. After modification, again a check is made for all the nutrients.

Exercise

1. A laying hen is consuming 100 g of feed containing 285kcal of metabolizable energy per day. Comment whether the diet is meeting the energy requirement as per the recommendation of ICAR (1985).
2. Formulate 'Broiler starter' diet each to contain conventional or unconventional feedstuffs.
3. Formulate 'Broiler finisher' diet each to have conventional or unconventional feedstuffs.
4. Formulate one 'Layer starter' (chick) diet using commonly available feedstuffs of your locality.
5. Formulate one 'Layer finisher' (grower) diet each using conventional or unconventional feedstuffs.
6. Formulate on 'Layer' diet each using conventional or unconventional feedstuffs.
7. Visit a commercial poultry farm and observe the feeding practices that are followed for feeding of various classes of poultry and note the salient features including feed formulations of that farm.

ANNEXURE – 1

Feed composition table for important poultry feeds (Dry matter basis).

Feedstuff	DM (%)	ME (Kcal/kg)	Protein (%)	Ca (%)	P (%)
Blood meal	88.8	1420	73.4	0.32	0.31
Bone meal	95.5	1044	14.6	27.00	12.11
Fat	97.9	7700	--	--	--
Meat meal	92.5	2319	56.2	2.68	2.06
Meat and bone meal	94.0	2111	53.8	11.25	5.39
Liver residue meal	90.9	3000	65.4	0.54	1.35
Bajra grain	89.6	2642	12.7	0.13	0.72
Barley grain	96.4	2618	12.0	0.29	0.63
Coconut meal	91.0	1190	22.6	0.23	0.66
Fish meal	93.8	1834	43.1	7.16	1.67
Groundnut meal (deoiled)	93.3	2328	48.6	0.31	0.67
Guar meal	89.7	1893	42.0	0.54	0.70
Jowar grain	87.3	2645	10.3	0.18	0.32
Linseed meal	90.7	1671	29.6	0.48	0.98
Lucerne leaf meal (dehydrated)	91.0	1777	19.1	1.83	0.45
Maize grain	89.5	3309	9.2	0.25	0.40
Maize gluten mel	90.3	2705	49.9	0.22	0.35
Mahua cake	90.0	2169	19.4	0.20	0.80
Mustard meal	91.3	2373	35.1	0.89	1.78
Oats grain	91.7	2848	14.7	0.11	0.41
Poultry byproduct meal	93.0	--	56.4	3.95	1.73
Feather meal	91.0	--	84.6	0.65	0.32
Poultry manure (dehydrated)	93.5	791	25.7	3.86	1.42
Rice polish	91.8	2937	12.7	0.27	1.37
Deoiled rice bran	92.3	2235	14.1	0.37	1.80
Safflower meal (decorticated)	90.0	2039	42.0	0.40	1.25
Safflower-meal (undecorticated)	95.0	1498	22.0	0.34	0.84
Sal seed meal, deoiled	90.4	3096	10.4	0.24	0.16
Sesame meal	90.7	1882	39.1	2.46	1.42
Silkworm pupae meal (deoiled)	90.5	3000	69.8	0.29	0.58
Soybean meal	89.9	2694	41.7	0.36	0.90
Molasses	73.6	2400	2.8	1.51	0.66
Sunflower meal	89.1	2230	37.2	0.43	1.14
Tapioca flour	90.0	3000	2.9	0.58	0.12
Tapioca waste	90.4	2339	4.1	0.58	0.19
Wheat bran	88.9	1069	14.7	0.19	1.12
Wheat grain	89.8	3045	10.3	0.18	0.43
Yeast sludge	93.0	1848	45.0	0.10	0.41

16. Methods for Improving Nutritive Quality of Straws and Other Crop Residues

Crop residues and other by-products of agriculture has become the mainstay for the nutrition of ruminant animals because the availability of fodder from uncultivable lands and other areas is becoming highly limited. There is an abundant supply of crop residues, particularly cereal straws during the dry periods because the dry season often coincides with the harvesting time of cereal crops in our country.

Most roughage, particularly crop residues and agro-industrial by-products are low in digestibility and deficient in nitrogen, minerals and vitamins. They are also physically resistant to combination and the high amounts of indigestible lignin and silica contained in them limits their intake. Hence these roughages alone cannot sustain effective animal production. With most poor quality roughages animals may consume DM at 60-80 g/kg $W^{0.75}$ /day; an amount hardly enough even to support maintenance requirement before poor quality roughages are considered satisfactory ruminant feedstuffs.

Methods of Improving the Feeding Value of Crop Residues

The objective of designing a diet based on poor quality roughages should be to optimize animal productivity rather than to meet requirements for a given production process. In this context, it is important to consider the following criteria:

1. Establishment of efficient rumen fermentation in order to optimize microbial growth and maximize digestibility of the fibre component of the diet within the rumen.

2. Provision of dietary nutrients to complement and balance the requirements for a particular productive process by supplementation with protein, energy and minerals.

Optimizing animal productivity using poor quality roughages can be achieved either by supplementation and or through pre-treatment.

A. *Supplementation* - The purpose of supplementing poor quality roughage is to correct nutrient imbalances and thereby create optimum rumen conditions for efficient microbial fermentation.

B. Treatment - The main objective of treating poor quality roughage is to increase its digestibility and or voluntary intake so as to increase the intake of digestible energy. Apart from supplementation many new methods have been developed to increase the efficiency of utilization of roughage material with the aim of achieving the above objectives. Figure 1 summarizes the methods available for this purpose.

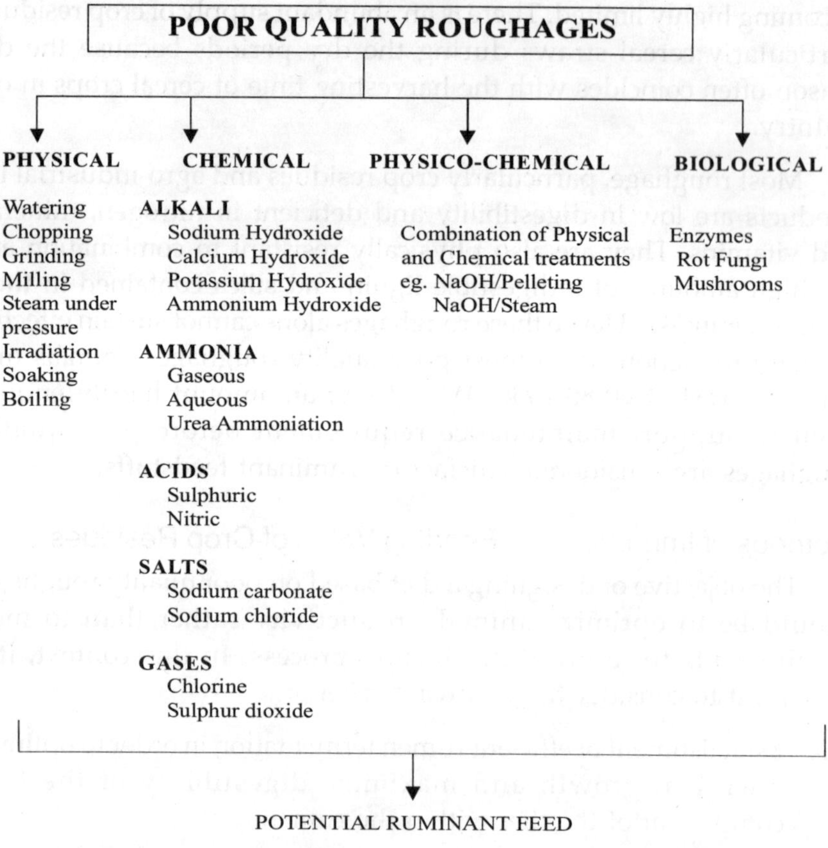

Fig. 1 : Methods of improving the nutritive value of poor quality roughages.

Methods of Improving the Nutritive Value of Poor Quality Roughages

Alkali Treatment : Most attention has been given to sodium hydroxide (NaOH) but similar improvements in digestibility have been reported with calcium hydroxide and ammonium hydroxide although treatment conditions may be more demanding. Other alkalis such as potassium hydroxide & sodium carbonate have also been investigated.

Methods of Treatment: Alkali treatment can be of two major types:

Wet method: This includes 1. Beckmann method, 2. Modified Beckmann method or Torgrimby method, 3. Dip treatment.

Dry method: 1. Spray treatment, eg. Small scale or on farm treatment, Industrial or commercial scale treatment.

Wet Methods

Wet methods of treatment involve soaking of the roughage in a chemical solution as in the Beckmann method (Beckmann, 1921).

1. The Beckmann's method, the original wet method of treatment consists of treating chopped straw in 8-10 times its weight of cold 1.2-1.5% (w/v) solution of NaOH for at least 4 hours. The treated straw was drained and washed with a large quantity of water until free from alkali. The quantity of water used to wash the treated straw varied from 40-50% of the added NaOH and was generally reused after charging with more NaOH. The treatment dissolved 20-25% of dry matter from the straw and therefore, was lost. The feed so prepared had a starch equivalent of 56. About 65 % of nitrogen free extractives and 85% crude fibre were utilized by animals.

   Although the method had many advantages compared to the older boiling methods and inspite of promising results with feeding trials, in the light of the present knowledge the whole procedure was very laborious and has got very limited application practically.

2. The Torgrimsby method (modified Beckmann method), an improvement over the Beckmann method is a closed treatment system involving little NaOH (0.75 g/kg straw). The amount of water added to the system is equal to the amount of water removed in the treated straw. The easiest way to visualize the method is to follow the daily sequence on a small farm. Three tanks are needed, each with an attached drain board. Tank A contains 1000 litres of a 1.5% NaOH solution in which 100 kg of straw, in bundles or bales, are soaked. The first washing tank B is twice as long as A and contains 2000 litres of water. The second washing tank C, the same size as A, contains 1000 litres of water.

Step	Time	Operation
1	07.00	Remove treated straw from A and place on board to drain. This straw was placed in A the day before at 12.00 noon.
2.	08.00	Remove drained treated straw from board and place in B. Place fresh (dry) straw (100kg) in B as well.
3.	12.00	Make up NaOH concentration of A by adding 4 kg NaOH. Remove fresh straw from B and place in A. Remove treated straw from B and place on board to drain.
4.	13.00	Remove treated straw from board and place in C.
5.	16.00	Remove treated straw from C and place on board to drain
6.	17.00	Transfer 300 litres of water from C to B to make up the volume of the latter (100kg fresh straw removed 300 litres of water earlier).
7.	18.00	Wash treated straw on drain board of C with 300 litres of fresh water by pouring over the straw. This runs into C, making up its volume. The treated straw is now ready for feeding.

Straw treated by the Torgrimsby method has an in vitro dry matter digestibility of about 70 per cent, an increase of 32 per cent over the untreated digestibility of 38 per cent. The method has the following advantages:

- It uses less NaOH than the Beckmann method;
- It uses less water than the Beckmann method;
- Dry matter loss is reduced;
- There is no pollution because it is a closed system.

3. The dip treatment was developed by Sundstol and co-workers in Norway and Tanzania. To avoid excessive amounts of NaOH in the final feed soaking time may be reduced to 2h or even less than 1h. After the draining of excess alkali solution the straw is allowed to 'ripen' for 3-6 days as there is evidence to show that digestibility of treated straw increases during the 'ripening' process.

Dry Methods

1. With a view to eliminating the disadvantages of the Beckmann system, Wilson and Pigden (1964) evolved a dry process of treatment in which straw is treated with a small volume of concentrated solution of NaOH. The straw is sprayed or sprinkled with NaOH while being mixed. Research has shown that 4 to 6 kg of NaOH dissolved in 200 litres of water is adequate to wet 100 kg of straw. The quantity of solution required is less (100-120 litres) if a pressure sprayer is used.

Straw treated with NaOH in this manner is moist and has a pleasant odour. Animal consume treated straw readily, often resulting in 30-40% increase in intake compared to untreated straw. Digestibility is generally increased by 10-15 percentage units raising the ME value of straw from about 5.5 to 6.0 to 8.0 to 9.0 MJ/kg of dry matter. In practical terms this would mean an improvement in straw quality to be equivalent to medium quality hay. Alkali treated roughages have been found to improve animal performance.

The achievements from the use of NaOH to upgrade poor-quality roughages are diminished by some draw backs such as high price of sodium hydroxide, the corrosive nature of the chemical and the necessity for protective equipment and trained personnel for application. It has also been suggested that feeds containing unreacted alkali and/or high Na^+ levels might inhibit rumen fermentation. Such disadvantages have made sodium hydroxide treatment unfavorable amongst many farmers.

Ammonia Treatment

Ammonia is an alternative to sodium hydroxide for treatment of poor quality roughages. The advantages are: (1) It does not leave residual alkali as NaOH and (2) It increases the nitrogen content of the treated material by 0.8 to 1.0 percentage units. Under Indian conditions Urea-ammoniation treatment is the most practical method of treating the fibrous crop residues.

Urea-Ammoniation Treatment

One of the most promising methods of ammoniation is to use fertilizer grade urea as a solution (4-5% W/W) and preserve the sprayed material under air tight conditions for 3-4 weeks. In the presence of moisture and enzyme urease, urea is decomposed to form ammonia and carbondioxide. Since urease enzyme is present in most crop residues, using urea to generate ammonia *in situ* a very important process for treatment of poor quality roughages.

$$NH_2\overset{\overset{\displaystyle O}{\|}}{C}NH_2 + H_2O \longrightarrow 2NH_3 + CO_2$$

The following steps are involved in treating 100g of crop residue in the laboratory.

Materials Required

100g of crop residue (paddy/jowar straw), tray, urea, water, weighing scale, measuring cylinder, polythene bags or bottles for storage.

Procedure

1. The following are the standard conditions required for treatment of various crop residues.

Crop residue	Quantity of urea Required (kg/100kg straw)	Level of moisture required (%)	Incubation period (days)
Paddy straw	4	85	10
Jowar straw	4	50	14
Maize straw	4	50	20
Bajra straw	4	35	20
Korra straw	4	30	30
Ragi straw	4	100	20

2. The quantity of urea mentioned in the above table should be dissolved in the water calculated as per the moisture required e.g. 4g of urea to be dissolved in 85ml of water (to get 50% moisture level).
3. Weigh 100g of crop straw.
4. Sprinkle the dissolved urea on the straw and mix the straw thoroughly.
5. Fill the treated material into a polythene bottle and close the lid (air) tightly.

6. Store the material for 20 days to allow the released ammonia to react.
7. After 20 days of incubation at room temperature, open and observe the following and record the observations.
   i. The smell of ammonia upon spreading the material in a tray.
   ii. The colour of the treated straw against untreated.

Exercise

1. Calculate the quantity of water to be used to dissolve urea and sprinkle to bring to 50% moisture level, if the paddy straw contains 80% dry matter.
2. Closely observe the demonstration of urea-ammoniation of paddy straw at laboratory scale and practice at farm scale by treating 100kg of paddy straw and observe its quality after the completion of the incubation period.

17. Utilization of Some Conventional and Unconventional Feedstuffs for Livestock Feeding

Ruminants generally receive their major nutrient requirements through grazing in the natural grasslands which have hitherto been mostly unsystematically exploited (both by overgrazing and under grazing). The pattern of production on range grazing consists of good production (growth, milk, wool etc.) during wet period, coinciding with good herbage growth, followed by variable decline in the production during the dry period in which herbage growth is affected. To sustain optimum production, roughages and concentrates depending on their availability, are given as supplemental feed during the lean dry periods.

In animal production system, conventional feeds like forages, cereals, milling byproducts and oil cakes are given to both ruminants and non-ruminants. These feeding systems are mostly practiced in the temperate regions where agriculture is traditionally much advanced. In India their place is taken by the agro-industrial byproducts, which include both the crop residues (paddy straw, wheat straw, stovers etc.) and unconventional feeds like sal seed meal, palm kernel cake etc. The term non-conventional feed (NCF) or non-conventional feed resources (NCFR) or unconventional feed (UCF) is a relative term and may differ from country to country and region to region, and time to time in the same country. Normally, NCF refers to the feed ingredient which has hitherto not been used for animal feeding. For example, paddy straw is a conventional feed in India, as it is being fed to cattle and buffaloes for centuries but in many parts of the world it is a non-conventional feed as it is not fed to the animals because of the availability of good grasses through pasture and other fodder crops. The NCF is generally organic and is the end product of a particular production system. It could be bulky (crop residues) or concentrate (seed meal etc) supplement.

Table: Nutritive value and level of inclusion of most commonly used unconventional feed ingredients in livestock rations.

Name of feed Ingredients	DCP	TDN	Level of inclusion	Presence of incriminating factors	Availability in India (10^5MT)
Sal seed meal	0.1	57.8	10% in CM of milch cattle 25% in calf starters	Tannic acid 8-10%	0.70
Tamarind seed waste (Decorticated)	11.26	63.92		Nil	87.0
Rubber seed cake	18.6	66.0	25-30% in CM of cross bred calves and milch cattle	HCN (9mg/100g)	1.5
Babul seeds	13.8	59.0	30% in CM of crossbred calves	Tannins (5%)	0.60
Mango seed kernels	6.1	70.0	10% in CM of milch cattle	Tannins (5-6%)	10
Cassava starch waste	1.8	64.0	25% in CM of crossbred calves	NIL	0.40
Apple waste	2.0	70.0	30% in CM of crossbred calves and 12.5% level in broiler mash	NIL	0.20
Ambadi cake	18.7	63.8	20% level in CM of cross bred calves	NIL	0.30
Castor bean meal	--	--	Detoxified meal at 10% in CM of sheep	Ricin (0.22%)	0.60
Karanj cake	25.5	62.0	15% in CM of crossbred calves and lactating cows.	Karanjine & some unidentified factors (10-15mg/100 g of cake)	1.3
Kosum cake	14.73	79.62	35% in the CM of crossbred calves. 17% in broiler mash	HCN (2.4 mg/100g)	0.30
Mahua seed cake	9.3	49.80	20% level in the CM of crossbred calves.	Mowrin 19%	3.0
Niger seed cake	32.7	49.4	57% level in CM of crossbred calves	NIL	1.0

18. Feed Conversion Efficiency

There are various ways by which the efficiency of feed conversion into animal products can be determined, namely, conversion of dry matter, energy and protein in to animal products.

i. *Dry Matter* - The amount of feed required to produce a unit of live weight gain milk production, egg production etc. is determined. For instance to produce 1 kg live weight gain for the broilers about 2kg of mash is required therefore the feed efficiency is 2:1. Likewise in the case of pigs the amount of feed required to produce 1kg live weight gain is about 4 kg up to a slaughter weight of 70kg. This procedure is useful where various diets are being compared with regard to efficiency for a particular biological function but when relative units of conversion of feed into various products in different species are considered it is not a good method. The types of feed given to various animals are different in nutritional characteristics. For instance in ruminants fodders, straws etc. are fed which can not be consumed by the pigs, poultry and human beings.

There are many factors which influence the cost benefit ratio like the price of the ingredients, labour, energy, production levels, socio-economic factors like product preference, its cost etc. Therefore, no single yardstick can be applied for a region as it varies from country to country and from time to time in the same country. Ranjhan estimated the efficiency of feed conversion for different traits of buffaloes in India which is presented below:

Table : Efficiency of feed conversion for different production traits in Riverine and Swamp buffaloes.

	Body weight (kg)	Trait	Level of production	Feed conversion kg/unit production
Buffalo (Riverine)	500	Milk	2500 l	9 : 1[a]
Buffalo (Swamp)	400	Milk	1000 l	10 : 1[b]
Buffalo (Riverine)	350	Cara beef	36 m	9 : 1[a]
Buffalo (Swamp)	350	Cara beef	42 ml	10 : 1[b]
Buffalo (Riverine)	450	Draught	0.60 HP	4:1
Buffalo (Swamp)	400	Draught	0.75 HP	5:1

a, 50% roughage and 50% concentrate; b, 90% good quality roughage and 10% concentrate; c, exclusively on good quality roughage; m, months; HP horse power; l, litre.

Goats have been reported to digest fibre more efficiently and utilize nutrients better than cattle and buffaloes. Goat rumen liquor digests significantly more cellulose as compared to sheep. Calculations of the efficiency of meat and milk production in goats revealed that meat production efficiency was comparable to sheep and cattle, whereas, the efficiency of milk production was found to be some what higher in goats. Comparison of feed efficiency under intensive system of rearing revealed that goats required on an average 17 kg feed per kg gain as compared to 15 kg in sheep.

□□□

Part-II

HUMAN AND PET NUTRITION

1. Important Terms in Nutrition

Food

Food is the sourcre of energy and tissue building constituents.

Nutrition

Nutrition involves various chemical reactions and physiological processes which transform foods into body tissue and activities. It involves the ingestion, digestion and absorption of the various nutrients, their transport to all body cells and the removal of unusable elements and waste products of metabolism.

Nutrients

It is the substance that furnishes nourishment to the body. The nutrients are constituents of food which must be supplied to the body in suitable amount to maintain health. Nutrients include carbohydrates, proteins, fats, vitamins and minerals.

Nutritional Status

It is the condition of health of an individual as influenced by the utilization of nutrients.

Malnutrition

Malnutrition means a disorder in nutrition and disorder may be excessive nutrition or under nutrition.

Under Nutrition

It is the state of ill health which can be attributed to the deficiency of nutrients due to poor intake.

General under nutrition

It is the total reduction in the food intake as seen in varying degrees of starvation.

Partial under nutrition

This arises from a deficiency of one or several nutrients.

Balanced ration

The ration contains a combination of feed ingredients in order to provide all essential nutrients in proper amounts and proportions based on the standard recommendations for a particular species under specified physiological conditions such as growth, pregnancy or production where the quality and quantity of the products are also considered.

2. Principles and Planning of Balanced Diet Charts for Human

Food Groups

Various food items commonly consumed in our country are the main source of nutrients while formulating nutritionally adequate diets for various categories of people to meet their needs as per the nutritional standards. In order to do this conveniently food groups system is widely followed.

'Food group system' converts quantitative nutrient data into food related information that can be used both by consumer and health professionals in diet planning to achieve nutritional adequacy.

Food may be broadly classified into groups based on their nutritive value:

1. Cereals and millets
2. Pulses (legumes)
3. Nuts and oil seeds
4. Vegetables
5. Fruits
6. Milk and milk products
7. Eggs
8. Meat, fish and other animal foods
9. Fats and oils
10. Sugar and other carbohydrate foods and
11. Spices and condiments

Food can be placed into five groups depending upon the major nutrient. The five group plan allows a person to plan his/her diet to achieve the nutritional adequacy as per 'Recommended Dietary Allowances' (RDA). Information on foods which are rich sources of nutrients is given in Table 1.

Nutritional Classification of Foods

The above groups of foods may be broadly classified under 3 heads from nutritional point of view:

i. Energy yielding foods
ii. Body building foods and
iii. Protective food

Energy Yielding Foods

This group includes foods rich in carbohydrates and fat and also pure fats and carbohydrates. Cereals, roots and tubers, dried fruits, sugar and fats form the important energy yielding foods. Cereals however provide in addition to energy the greater part of the proteins, certain minerals and vitamins in the diets of the low income groups in the developing countries.

Table 1: Five food group system

Food Group	Main Nutrients
Cereals grains and Products Rice, Wheat, Ragi, Bajra, Maize, Jowar, Barley, Rice flakes, Wheat flour	Energy, Protein, Invisible fat Vitamin-B1, Vitamin-B2, Folic acid, Iron, Fibre
Pulses and Legumes Bengal gram, Black gram, Green gram, Red Gram, Lentil (whole as well as dhals), Cowpea, peas, Rajmah, Soybeans etc.	Energy, Protein, Invisible fat, Vitamin-B1, Vitamin-B2, Folic Acid, Calcium, Iron, Fibre
Milk and meat Products	
Milk: Milk, Curd, Skimmed milk, cheese	Protein, Fat, Vitamin-B2, Calcium
Meat: Chicken, Liver, Fish, Egg, Meat	Protein, Fat, Vitamin-B2
Fruits and Vegetables	
Fruits: Mango, Guava, Tomato, Ripe, Papaya, Orange, Sweet lime, Water melon	Carotenoids, Vitamin-C, Fibre
Vegetables (Green leafy) Amaranth, Spinach, Gogu, Drumstick leaves, Coriander leaves, Mustard leaves, Fenugreek leaves.	Invisible fats, Carotenoids, Vitamin-B2, Folic Acid, Calcium, Iron, Fibre
Other Vegetables: Carrot, Brinjal, Ladies fingers, Capsicum, Beans, Onion, Drumstick, Cauliflower	Carotenoids, Folic Acid, Calcium, Fibre
Fats and Sugars	
Fats: Butter, Ghee, Hydrogenated oils, Cooking oils like groundnut, Mustard, Coconut	Energy, Fat, Essential Fatty acids
Sugar: Sugar, Jaggery	Energy

Body Building Foods

Foods rich in proteins are called body building foods. This may be broadly divided into two groups: (a) Milk, egg, meat and fish which are rich in proteins of high biological value and (b) Pulses, oil seeds and nuts and low fat oil seeds flours which are rich in proteins of medium nutritive value.

Protective Foods

Foods rich in proteins, vitamins and minerals are termed as protective foods. Protective foods are broadly classified into two groups: (a) Foods rich in vitamins, minerals and proteins of high biological value, eg: milk, eggs, fish and liver and (b) Foods rich in certain vitamins and minerals only, eg. green leafy vegetables and some fruits. The nine food group plan is given in Table 2.

Table 2 : Nutritional classification of foods (Nine food group plan)

Food Group	Main Nutrients contributed
Protective foods	
Group I Milk	Proteins, Calcium, Vitamin A, Riboflavin and Vitamin B12
Liver	Proteins, Vitamin A, Riboflavin, Folic acid, Vitamin B12 and Iron
Eggs	Proteins, Vitamin A, Riboflavin, Vitamin B12 and Iron
Group II	
Green leafy vegetables	Pro-vitamin A (carotene), calcium, iron, folic acid, riboflavin and vitamin C.
Group III Fruits	Fair to good sources of vitamin C.
Body Building foods (Protein rich foods)	
Group IV	
Milk, meat, fish, eggs and liver	Proteins of high biological value and vitamins and minerals listed above
Group V	
Pulses, nuts & oil seeds (except coconut)	Proteins of moderate nutritive value, thiamine, riboflavin and iron
Energy yielding foods	Supply more than 70% calories, more than50% B-vitamins and more than 7% proteins in poor diets.
Groups VI	
Cereals (except ragi)	Supplies more than 6% proteins, more than 50% B-vitamins, more than 80% calcium and 70% calories in poor ragi diets.
Ragi	Good sources of starch and fair sources of Vit. C
Group VII	
Roots and tubers (potato, sweet potato and tapioca)	Rich source of provitamin A
Carrots	Rich sources of carbohydrates
Group VIII	
Sago, sugar and jaggery	Rich sources of essential fatty acids and Vitamin E
Group IX	
(oils and fats)	
Vegetables oils and fats (except coconut oil)	Good sources of Vitamin A
Vanaspati, butter and ghee	

Planning of Balanced Diets

A balanced diet may be defined as one which contains various groups of food stuffs such as energy yielding, body building and protective foods in correct proportion so that an individual is assured of obtaining the minimum requirements of all the nutrients. The components of a balanced diet differ according to age, sex, physical activity, economic status and physiological state, such as pregnancy, lactation etc.

Balanced Diets at High Cost

These diets include liberal amounts of protective and protein-rich foods such as milk, eggs, meat, fish and fruits and moderate quantities of cereals, pulses, nuts and fats.

Balanced Diets at Moderate Cost

These diets include moderate amounts of protective and protein rich foods such as milk, eggs, meat, fish, fruits and fats and liberal amounts of cereals, pulses, nuts and green leafy vegetables.

Balanced Diets at Low Cost

These diets include minimum amounts of protective and protein rich foods such as milk, eggs, meat, fish and fats and liberal amounts of cereals, pulses, nuts and green leafy vegetables.

Ill Balanced Diets

An ill balanced diet is deficient in calories, proteins, essential vitamins and minerals as it is lacking in protein-rich and protective foods.

Balanced diets at varying costs for pre-school children as compared with an ill balanced diet are given in Table 3.

Table 3 : Balanced diets at varying costs for a pre-school child aged 5 years (g/ caput/day) compared with ill balanced diet.

Food stuffs	Ill balanced diets	Low cost		Moderate cost		High cost	
		V	NV	V	NV	V	NV
Energy yielding foods							
Cereals	150	150	150	120	120	100	100
Fats	5	15	15	20	20	25	25
Sugar and Jaggery	20	40	40	40	40	40	40
Body building foods							
Pulses	10	60	45	40	30	40	30
Groundnut (roasted)	10	30	30	--	--	--	--
Meat, fish and eggs	--	--	15	--	30	--	50
Protective foods							
Milk	80	250	200	600	500	1000	800
Leafy vegetables	20	75	75	50	50	50	50
Fruits and fruit juice	10	50	50	100	100	200	200
Other foods							
Other vegetables	30	50	50	50	50	50	50

V- Vegetarian NV – Non Vegetarian

ICMR Balanced Diets

The dietary allowances for Indians were first recommended in 1944 by the Nutrition Advisory Group. These recommendations were revised in respect of energy and protein in 1958. The Nutrition Advisory committee of the Indian Council of Medical Research has recommended dietary allowances for Indians in 1968. Recommended dietary allowances have been updated by the 1978 committee of ICMR. A new set of balanced diets based on the latest recommended dietary allowances (1988) have been formulated. In these formulations some of the shortcomings of the earlier balanced diets have been overcome. In evolving these diets cost factor has received adequate consideration. There least cost balanced diets have been formulated by using linear programming at National Institute of Nutrition, after using appropriate constraints consistent with practical considerations. The constraints used in developing the present diets are as follows:

1. Energy derived from cereals to be not more than 75% of total requirements.
2. Ratio of cereal protein to pulse protein has been kept within the range of 4:1 and 5:1.
3. Minimum level of leafy vegetable has been used
4. Quantity of other vegetables has been kept at 80g or less and all vegetables put together to be not more than 150g.
5. A minimum milk intake of 100ml.
6. Energy derived from fat or oil not to exceed 15% of total calories.
7. Energy derived from refined carbohydrates (sugar or jaggery) has been kept around 5% and total calories from fat and sugar not to exceed 20%.
8. Levels of food items suggested should be consumable by an average individual.

No separate diets have been suggested for non-vegetarians, but replacement of pulses by suitable quantities of egg, fish or meat has been suggested (from 50 to 100%). For lactating and pregnant women, additional allowances of cereals, pulses, milk and fat have been suggested to meet the additional requirement of various nutrients.

The recommended balanced diets, additional suggestions for non-vegetarian diets and suggested additional allowances for pregnancy and lactation are given in tables 4, 5 and 6, respectively.

Table 4 : Balanced Diets (ICMR 1981) (Food stuffs-g/day/caput)

Age group	Cereals	Pulses	Leafy vegetable les	Other vegetable les	Roots and tubers	Milk	Oils & fats	Sugar & jaggery
Adult Men								
Sedentary	460	40	40	60	50	150	40	30
Moderate work	520	50	40	70	60	200	45	35
Heavy work	670	60	40	80	80	250	65	55
Adult Women								
Sedentary	410	40	100	40	50	100	20	20
Moderate work	440	45	100	40	50	150	25	20
Heavy work	575	50	50	100	60	200	40	40
Pregnancy	+35	+15	--	--	--	+100	--	+10
Lactation	+60	+30	--	--	--	+100	+10	+10
Children								
1-3 years	175	35	40	20	10	300	15	30
4-6 years	270	35	50	30	20	250	25	40
Boys(10-12)	420	45	50	50	30	250	40	45
Girls (10-12)	380	45	50	50	30	250	45	45

+ Additional allowances during pregnancy and lactation.

Table 5 : Suggested substitution for Non-vegetarians

Food item which can be deleted from non-vegetarian diets	Substitution that can be suggested for deleted item or items
50% of pulses (20-30g)	1. One egg or 30g meat or fish 2. Additional 5g of fat or oil
100% of pulses (40-60g)	1. Two eggs or 50g meat or fish or one egg + 30g meat or fish 2. 10g of fat or oil

Table 6 : Additional allowances during pregnancy and lactation

Food items	During pregnancy	Calories (Kcal)	During lactation	Calories (Kcal)
Cereals	35g	118	60g	203
Pulses	15g	52	30g	105
Milk	100g	83	100g	83
Fat	--	--	10g	90
Sugar	10g	40	10g	40
Total		293		521

Recommended Dietary Allowances for Indians

The Nutrition Advisory Committee of Indian Research Fund Association, now Indian Council of Medical Research recommended for the first time, dietary allowances of Indian energy, proteins, iron, calcium, Vit. A, thiamine, ascorbic acid and Vitamin D. These recommendations were advised in 1958 in respect of energy and protein. The nutrition Expert Group of Indian Council of Medical Research revised in 1968 the earlier recommendations of nutrient allowances for Indians in respect of all nutrients except energy. The committee also considered the desirable level of intake of additional nutrients like folic acid, Vit. B12, vitamin D and fat. In making these recommendations, the committee was guided by the safe allowances of various nutrients suggested by the expert groups of the Food and Agricultural Organizations and the World Health Organization and data on nutrient requirement of Indians. These recommendations were again revised in 1978 by an Expert Group of the ICMR and allowances were updated and revised in respect of several nutrients. Nutrient allowances for Indians were again examined in detail by an Expert Committee of the ICMR in April, 1988 and detailed recommendations with respect to safe levels of intake of all nutrients were made. The basis of these recommendations is discussed in detail in the Report of the Committee. Here the recommended intakes are given in Table 7. The main point on which they are based and guidelines to use them are described below briefly.

1. Body Weight

Reference man and women: On the basis of height of Indian adult, body weight of 60kg and 50kg for reference man and reference woman should be used in arriving at RDA, instead of 55kg and 40kg used earlier.

2. Energy Allowances

a. Energy requirement is expressed in terms of BMR values which are derived from the body weight of Indians using the equation given by FAO, 1980 and the values reduced by 5% to allow for the lower BMR of Indians.

b. Energy allowance for children and adolescents are for normally growing healthy Indian children. Allowance is made for the age and no correction for actual body weight of children needs to be made.

3. Protein Allowances

Protein allowances are given in terms of mixed vegetables proteins contained in Indian diets, the NPU of which is assumed to be 65. Adult protein requirement is 1g/kg.

4. Fat Requirement

Fat requirement was examined in detail and more realistic estimates of minimum intakes have been given. In arriving at fat requirement, the total invisible fat content of cereal based diets eaten in India and the minimum EFA requirement of various groups were considered.

5. Minerals

Iron requirements have been arrived at using the factorial approach taking into account the basal loss in case of men, basal loss+ growth requirement in case of children. Dietary iron requirements are given at levels of absorption of 5%, 3% and 2%.

Applicable respectively to the diet based on rice, based on mixed cereals (rice+wheat) and based wholly on wheat or millets.

Calcium and phosphorus : Because of close relationship between Ca and P, allowances for both these nutrients are given. The desirable ratio of Ca: P is 1:1.5 in case of infants and 1:1 in case of other groups.

Trace elements : Desirable daily intakes of some trace elements have been suggested for an adult: chromium 65 μg, copper 2.2 mg, manganese 5.5 mg, zinc 15.5 mg and molybdenum 500 μg.

6. Vitamins

Vitamin A: The recommended level of vitamin A (retinol) for adult is 600 μg. In terms of beta-carotene it will be 2400 μg.

Thiamine, Riboflavin and Nicotinic Acid

The daily allowances for these vitamins are related to energy intake. The basic allowances per 1000 kcal are thiamine 0.5 mg, riboflavin 0.6 mg and nicotinic acid 6.6 mg niacin equivalent. Niacin allowance takes into account the contribution of trytophan also, assuming 60mg dietary tryptophan yields 1 mg of niacin in the body.

Table 7: Recommended Dietary allowances for Indians

Group	Particulars	Body wt. Kg	Net energy Kcal/d	Protein g/d	Fat g/d	Calcium mg/d	Iron mg/d	Vit. A.µg/d Retinol	Vit. A.µg/d β-carotene	Thiamin	Riboflavin	Nicotonic acid	Pyridoxin	Ascorbic acid	Folic Acid	Vit B-12
Man	Sedentary work		2425							1.2	1.4	16				
	Moderate work	60	2875	60	20	400	28	600	2400	1.4	1.6	18	2.0	40	100	1
	Heavy work		3800							1.6	1.9	21				
Woman	Sedentary work		1875							0.9	1.1	12				
	Moderate work	50	2225	50	20	400	30	600	2400	1.1	1.3	14	2.0	40	100	1
	Heavy work		2925							1.2	1.5	16				
	Pregnant woman	50	+300	+15	30	1000	38	600	2400	+0.2	+0.2	+2	2.5	40	400	1
	Lactation															
	0-6 months	50	+550	+25	45	1000	30	950	3800	+0.3	+0.3	+4	2.5	80	150	1.5
	6-12months		+400	+18						+0.2	+0.2	+3				
Infants	0-6 months	5.4	108/kg	2.05/kg		500				55 µg/kg	65 µg/kg	710 µg/kg	0.1	25	25	0.2
	6-12months	8.6	98/kg	1.65/kg				350	1200	50 µg/kg	60 µg/kg	650 µg/kg	0.4			
Children	1-3 years	12.2	1240	22			12	400	1600	0.6	0.7	8	0.9		30	
	4-6 years	19.0	1690	30	25	400	18	400		0.9	1.0	11		40	40	0.2-1.0
	7-9 years	26.9	1950	41			26	600	2400	1.0	1.2	13	1.6		60	
Boys	10-12 years	35.4	2190	54	22	600	34	600	2400	1.1	1.3	15	1.6	40	70	0.2-1.0
Girls	10-12 years	31.5	1970	57			19			1.0	1.2	13				
Boys	13-15 years	47.8	2450	70			41	600	2400	1.2	1.5	16	2.0	40	100	0.2-1.0
Girls	13-15 years	46.7	2060	65	22	600	28			1.0	1.2	14				
Boys	16-18 years	57.1	2640	78			50	600	2400	1.3	1.6	17	2.0	40	100	0.2-1.0
Girls	16-18 years	49.9	2060	63	22	500	30			1.0	1.2	14				

$$\text{Niacin equivalents} = \frac{\text{Niacin content} + \text{Trytophan content}}{60} \text{ (mg)}$$

Folic Acid

Allowance for folic acid in terms of free folate (*L. casei* activity) present in foods. RDA of folate in pregnancy will be 300 μg in addition to normal requirement of 100 μg since such high level of folate cannot be obtained. Therapeutic supplementation becomes necessary in pregnancy. Folic acid requirement can be expressed on the basis of body weight also i.e. 3 μg/kg.

Vitamin B12: Recommended intake of B12 is 0.2 - 1 μg/day. It is entirely derived from animal foods. This provides justification of including some animal foods in our daily diet.

Vitamin C : Earlier recommendations of 40 mg were retained. In case of infants, the RDA was changed from 20 to 25 mg/day.

Vitamin D : No dietary recommendations for vit. D was made since adequate vitamin D can be obtained through exposure to sunlight. In case of conditions of inadequacy of vitamin D medicinal supplements can be given.

Vitamin E : An intake of 0.8mg of vit. E per g of polyunsaturated fatty acid was suggested.

Vitamin B6: Earlier (1978) recommendations were retained.

Protein Calorie Ratio of Diets

It is useful to consider together the protein and energy requirement on habitual Indian diets. The protein requirement can be expressed as the ratio of protein calories to total dietary calories (PE %). A comparison of this PE% value with the protein energy% of the diet will indicate whether diets can fulfill protein needs if an adequate quantity of the diet is eaten to meet the energy requirement. The concept is useful because in many population groups enough diet is not consumed to meet energy needs, resulting in energy deficits. The ratio of protein requirement, expressed as the ratio of protein calories to the energy requirements is given in Table 8 for various population categories. It will be seen that a PE% between 8 and 12 would meet the protein requirement of any group provided its energy needs are met.

Table 8 : Protein Energy Ratios

Group	Recommended protein intake g/day	Recommended energy intake Kcal/day	Protein energy ratio (%)
Reference adult (Moderate activity)			
Man	60	2900	8.3
Woman	50	2200	9.1
Pregnant woman	65	2500	10.4
Lactating woman (0-6m)	75	2750	10.9
Infants			
0-6 months	2.0/kg	118/kg	--
6-12 months	1.7/kg	108/kg	--
Pre school			
1-3 years	21	1240	6.8
4-6 years	29	1690	6.9
7-9 years	40	1950	8.2
Adolescents			
13-15 yrs Boys	67	2450	10.9
Girls	62	2060	12.0
16-18 yrs Boys	75	2640	11.4
Girls	60	2060	11.7

Calcium Requirement

Calcium requirement has been measured by long term balance studies. Such studies among Western population, whose habitual diets contain high levels of Ca from generous amounts of milk, have indicated a requirement of Ca of the order 1 gm. Population groups in many developing countries subsist on a lower calcium intake of about 500mg without any ill effects. Long term, balance studies in such population indicate that they are in positive Ca balance even on much lower intakes. This is because the body can adapt to different levels of intakes of Ca and maintain a positive Ca balance. These studies were the basis for the RDAs for Ca suggested by the earlier committee which had given a range of intakes, as was also recommended by FAO/ WHO. However, the present committee after examining the evidence for Ca nutrition status of the Indian population suggested that the earlier RDA be retained. Instead of a range, a single figure for each group is suggested, as shown in table 10.

Table 9 : Indian foods rich in Calcium

Foods	Calcium mg/100g	Foods	Calcium mg/100g
Cereal/Millets		**Green Leafy Vegetables**	
Ragi	344	Agathi	1130
Amaranth seeds	510	Amaranth	397
Rajkeera seeds	223	Fenugreek	395
		Rape leaves	---
Pulses/Legumes		**Milk & Milk products**	
Bengal gram whole	202	Cow's milk	120
Horse gram whole	287	Buffalo milk	210
Rajmah	260	Cheese	790
		Khoa	Khoa
Nuts & Oilseeds		**Fish & sea foods**	
Gingelly seeds	1450	Hilsa	---
Mustard seeds	190	Rohu	---
Cumin seeds	1080		
Poppy seeds	1584		

Source: Nutritive value of Indian foods.

The committee also considered the desirable intake of phosphorus (P). Since P deficiency is unlikely on the types of diets consumed in India, ensuring an adequate P intake may not present a problem. It was suggested that an elemental Ca: P ratio of 1:1 may be maintained in most age groups, except in infancy where the ratio suggested is 1:1.5.

Table 10 : Suggested intakes of Calcium and Phosphorus (mg/d)

Group	Calcium	Phosphorus
Adult		
Men	400	400
Women	400	400
Pregnancy and lactation	1000	1000
Infants	500	750
Children		
1-9 years	400	400
10-15 years	600	600
16-18 years	500	500

Table 11 : Average Protein and energy values of common foods (per 100g of edible portion)

S.No.	Name of the food stuff	Protein (Nx6.25)g	Energy
Cereal Grains and Products			
1.	Bajra	11.3	361
2.	Barley	11.5	336
3.	Jowar	10.4	349
4.	Maize, dry	11.1	342
5.	Maize, tender	4.7	125
6.	Ragi	7.3	328
7.	Rice, parboiled, milled	6.4	346
8.	Rice, raw, milled	6.8	345
Cereal Grains and Products			
9.	Rice , flakes	6.6	346
10.	Wheat, flour (whole)	12.1	341
11.	Wheat, vermicelli	8.7	352
12.	Wheat, bread (brown)	8.8	244
Pulses and Legumes			
13.	Bengal gram, whole	17.1	360
14.	Black gram, dhal	24.0	347
15.	Green gram, dhal	24.5	340
16.	Peas green	7.2	93
17.	Rajmah	22.9	246
18.	Red gram, dhal	22.3	335
19.	Soybean	43.2	432
Leafy Vegetables			
20.	Amaranth caudatus	3.0	26
21.	Cabbage	1.8	27
22.	Cauliflower greens	5.9	66
23.	Coriander leaves	3.3	44
24.	Curry leaves	6.1	108
25.	Gouge	1.7	56
26.	Tamarind leaves, tender	5.8	115
Roots and Tubers			
27.	Arrowroot flour	0.2	334
28.	Beet root	1.7	43
29.	Carrot	0.9	48
30.	Mango ginger	1.1	53
31.	Onion big	1.2	50
32.	Onion small	1.8	59
33.	Potato	1.6	97
34.	Sweet potato	1.2	120
35.	Tapioca	0.7	157

contd...

Other Vegetables			
36.	Bitter gourd small	2.1	60
37.	Bottle gourd	0.2	12
38.	Brinjal	1.4	24
39.	Cauliflower	2.6	30
40.	Cucumber	0.4	13
41.	Drumstick	2.5	26
42.	French beans	1.7	26
43.	Ladies fingers	1.9	35
44.	Mango, green	0.7	44
45.	Snake gourd	0.5	18
46.	Tomato, green	1.9	23
Nuts and Oilseeds			
47.	Cashew nut	21.2	596
48.	Coconut fresh	4.5	444
49.	Groundnut roasted	26.2	570
Condiments and spices			
50.	Cardamom	10.2	229
51.	Chillies dry	15.9	246
52.	Chillies green	2.9	29
53.	Cloves dry	5.2	286
54.	Coriander	14.1	288
55.	Garlic dry	6.3	145
56.	Ginger fresh	2.3	67
57.	Pepper dry (black)	11.5	460
58.	Poppy seeds	21.7	408
59.	Tamarind pulp	3.1	283
60.	Turmeric	6.3	150
Fruits			
61.	Amla	0.5	58
62.	Apple	0.2	59
63.	Banana, ripe	1.2	116
64.	Grapes blue variety	1.3	37
65.	Grape fruit marsh's seedless	1.0	45
66.	Grape fruit triumph	0.7	32
67.	Guava, country	0.9	51
68.	Lemon	1.0	57
69.	Lime, sweet musambi	0.8	43
70.	Mango, ripe	0.6	74
71.	Melon, water	0.2	16
72.	Orange	0.7	48
73.	Palmyra fruit ripe (mesocarp)	0.7	87
74.	Palmyra fruit tender	0.6	29
75.	Papaya, ripe	0.6	32

contd...

76.	Pineapple	0.4	46
77.	Pomegranate	1.6	65
78.	Sapota	0.7	98
79.	Tomato, ripe	0.9	20
80.	Zizyphus	0.8	74
Fish and other Sea foods			
81.	Katla	19.5	111
82.	Prawn	19.1	89
83.	Rohu	16.6	97
Meat and Poultry			
84.	Buffalo meat	19.4	86
85.	Egg, duck	13.5	181
86.	Egg, hen	13.3	173
87.	Fowl	25.9	109
88.	Goat meat (lean)	21.4	118
89.	Liver goat	20.0	107
90.	Liver sheep	19.3	150
91.	Mutton, muscle	18.5	194
92.	Pork, muscle	18.7	114
Milk and Milk Products			
93.	Milk Buffaloes	4.3	117
94.	Milk cow's	3.2	67
95.	Milk human	1.1	65
96.	Curd (cow's milk)	3.1	60
97.	Butter milk	0.8	15
Fats and Edible oils			
98.	Butter	–	729
99.	Ghee (cow)	–	900
100.	Ghee (buffalo)	–	900
101.	Cooking oil (groundnut, gingelly, palmolein, mustard, coconut etc.)	–	900
Sugars			
102.	Sugar cane	0.1	398
103.	Honey	0.3	319
104.	Jaggery (cane)	0.4	383
105.	Sago	0.2	351
Beverages (Alcoholic)			
106.	Toddy, fermented	0.1	38
Beverages (Non-alcoholic)			
107.	Sugar cane juice	0.2	39

3. Preparation of Balanced Diet for New Born (Infants)

Human milk is universally regarded as the ideal food for feeding infants during the early months. Yet, due to various causes, the mother may not be able to feed the infant and feeding of infants with mammalian milk may be necessary. Cow's milk has been used from time immemorial for feeding infants in many countries. Buffalo and goats milk are being used, besides cow's milk, to a limited extent for feeding infants in some countries. The milk differ in composition for lactose, minerals and vitamins present in them. They also differ in their digestibility by the infants. During recent years dried milk foods are being used extensively for feeding infants. For successful feeding of infants, knowledge of nutritional requirements of infants and chemical composition of milk foods is necessary.

Nutritional Requirements of Infants

Data regarding the increase in body weight, length of infants, calorie and protein requirements of well fed infants are given in (Table 12). It is evident that normally growing infants double their weight at the 5th month and triple at 12 months of age.

Composition of Human, Cow, Buffalo and Goat Milk

The data regarding chemical composition of human milk as compared with those of cow, buffalo and goat milk are briefly discussed (Table 13).

Proteins

The Protein content of cow, buffalo and goat milk is about 3 times that of human milk. They contain larger amounts of casein which is responsible for the hard curd formed from these milk in the stomach of infants.

Table 12 : Mean body weight, Height, Calorie and Protein requirements of normal well fed Infants

Age (months)	Body length (cm) Male	Body length (cm) Female	Body Weight (kg) Male	Body Weight (kg) Female	Calories (Kcal)	Protein (g)
At Birth	47.5	47.0	3.1	3.0	370	6
1	50.6	50.0	3.8	3.7	430	7
2	53.7	53.0	4.5	4.4	480	8
3	56.8	56.0	5.2	5.1	520	9
4	58.8	57.9	5.8	5.6	560	10
5	60.8	59.8	6.4	6.1	600	11
6	62.8	61.7	7.0	6.6	640	12
7	64.5	63.4	7.4	7.1	680	13
8	66.2	65.1	7.8	7.5	720	14
9	68.9	66.8	8.0	7.7	760	15
10	70.2	68.1	8.2	7.9	800	16
11	71.5	69.4	8.4	8.1	850	17
12	72.8	70.7	8.6	8.3	900	18

Table 13 : Chemical composition of Human, Cow, Buffalo and Goat Milk (Values Per 100ml)

Nutrients	Human	Cow	Buffalo	Goat
Proteins (g)	1.2	3.3	3.8	3.3
Fat (g)	3.8	3.7	7.5	4.1
Lactose (g)	7.0	4.8	4.4	4.7
Calories (Kcal)	71	69	100	76
Ash (g)	0.21	0.72	0.8	0.77
Calcium (mg)	33	125	210	130
Phosphorus (mg)	15	96	130	106
Iron (mg)	0.15	0.10	0.20	0.05
Ascorbic acid (mg)	4.0	2.0	2.5	2.0

Fat

The fat content of human milk is nearly of the same order as those present in cow and goat milk, but buffalo milk contains about twice as much fat as that present in other milks.

Minerals

The iron content of human milk is nearly of the same order as those present in other milk. The contents of other minerals in cow, buffalo and goat milk are about 4-7 times those present in human milk.

Vitamins

Vitamin A and folic acid contents of human milk are nearly equal to those of other milks. Human milk contains higher amounts of ascorbic acid but lesser amounts of B-vitamins than those present in other milk.

Modification of Animal Milk for Infant Feeding

The milk of animals can be modified in the following ways for feeding infants.

1. Adjustment of Fat Content to that of Human Milk

This is needed only in the case of buffalo milk. Buffalo milk can be partially skimmed or toned with skimmed milk to have fat content of about 3.5%.

2. Boiling

Boiling brings about physico-chemical changes in the milk and helps to reduce markedly curd tension of milk from 36 to 50 units. It makes the milk wholesome.

3. Dilution with Water and Addition of Sugar

Animal milk is generally diluted with equal amounts of water to reduce the contents of proteins and various minerals for feeding infants of 1-2 months of age. The proportion of water is reduced gradually as the infant grows. Sugar is added at 5% level in the diluted milk. The quantities of milk, water and sugar required for feeding infants of various ages are given in Table 14.

Processed Infant Foods

Infant foods available on the market have widely varying protein (14-27%) and fat (10-25%) contents. They can be grouped broadly as followed: (1) whole milk food (full cream milk powder) with added vitamin D and iron. (2) Half cream milk food with added Vitamin D and iron and (3) Humanized milk food with added carbohydrate and protein with the later altered more or less to resemble those of human milk. An infant food with a protein content of 20-22% and fat content of 18% may be more suited for infants in tropical countries. "Amul Infant Food" is being manufactured as per the above standards and is marketed by the Kaira District Co-operative Milk Producers Union, Anand.

Most of the infant foods are fortified with vitamin D and some also with iron. They do not however, contain sufficient amounts of vitamin C. Hence artificially fed babies require a supplement of vitamin C. For feeding infants the proprietary infant foods are usually reconstituted with 8 to 10 times their weight of hot water depending on the age of the infant.

Supplementary Foods

Milk is deficient in iron and vitamin D and does not contain adequate amounts of ascorbic acid. When the infant is about 5-6 months old, strained baby foods based on vegetables, fruits and meat and precooked cereal foods are gradually introduced in the feed. These foods include the following:

(i) *Vitamin D:* A source of vitamin D is essential for infants receiving unfortified milk and milk foods. Great care should be taken to ensure that excess of vitamin D is not administered.

(ii) *Fruit juice:* Orange or tomato juice are commonly given to infants, which provide required amount of vitamin C. The organic acids presents in fruit juice help to lower the curd tension of milk and thereby aid in the digestion of milk proteins and in the absorption of calcium.

(iii) *Egg yolk:* Milk is a poor source of iron and a fair source of thiamine. Egg yolk is a fair source of both these nutrients and can be added to the diets of infants from 4th to 6th month onwards.

(iv) *Strained vegetables and fruits:* These are added to diets of infants from 4th month onwards. They provide iron, vitamin C and small amounts of indigestible carbohydrates. The later may help to prevent constipation.

(v) *Precooked enriched cereal foods:* Infants (aged 5-6 months) who receive small amounts of cereal foods thrive better than those who do not. Precooked roller dried cereal foods fortified with iron are used for this purpose.

Recently peanut protein isolate has also been used along with buffalo milk in the preparation of toned vegetable milk. This will help to compensate the milk shortage and to overcome malnutrition among infants.

Exercise

1. Calculate RDA for infants of 1, 3, 6 and 12 months of age and suggest a suitable feeding schedule to meet the nutrient requirement.

Table 14 : Feeding of Infants with fresh cow's or toned Buffalo Milk

	Age (months)							
	1	**2**	**3**	**4-5**	**6-7**	**8-9**	**10-11**	**12**
No. of feeds per day	8	7	6	6	6	5	5	5
Cow's milk or toned buffalo milk* (ml)	300	400	500	600-650	700-750	800-850	900-950	1000
Water (ml)	300	300	300	300	200	100	100	--
Cane Sugar (g)	30	40	50	60	70	70	70	70
Vitamin D (I.U.)	400	400	400	400	400	400	400	400
Egg yolk	--	--	--	--	½	½	1	1
Fruit juice (ml)	--	--	10	15	20	25	30	35
Cereal food (g)	--	--	--	10	15	25	30	40
Mashed vegetables and fruits (teaspoons)	--	--	--	--	4	4	6	8

• Dried milk foods reconstituted with 8 times the weight of water can be used.

4. Balanced Diets for Pre-School Children

The infant and the pre-school children are extremely vulnerable. They succumb readily to infectious diseases and infestations if the diet is poor in quality and quantity. The growth rate and development of infants and pre-school children depend on the adequacy of diet consumed by them. Frequent attacks of infectious diseases affect their growth and increase the requirements of various nutrients. The mean daily nutritional requirements recommended by ICMR Expert Group are given in Table 8 & 10. Balanced diets at high moderate and low costs for pre-school children are given in Table 15.

Balanced diets at high cost contain large amounts of milk, fruits, eggs, meat and fish which are costly, whereas the diets at moderate cost contain small amounts of milk and fair amounts of legumes and green leafy vegetables.

Table 15 : Balanced diets for Pre-school Children

FOOD STUFFS	High cost				Moderate cost				Low cost			
	1-3 yrs		4-6 yrs		1-3 yrs		4-6 yrs		1-3 yrs		4-6 yrs	
	V (g)	NV (g)	V (g)	NV (g)	V (g)	NV (g)	V (g)	NV (g)	V (g)	NV (g)	V (g)	NV (g)
Cereals	100	100	140	140	120	120	170	170	150	150	200	200
Pulses	30	20	40	30	50	40	60	50	50	40	60	50
Green leafy vegetables	50	50	75	75	50	50	75	75	50	50	75	75
Other Vegetables, roots and tubers	30	30	30	30	30	30	50	50	30	30	50	50
Fruits	100	100	100	100	100	100	100	100	50	50	50	50
Milk	1000	700	100	700	600	400	600	400	300	200	250	200
Egg	--	30	--	30	--	--	--	--	--	--	--	--
Meat and Fish	--	20	--	30	--	40	--	50	--	30	--	30
Fats and oils	20	20	25	30	20	20	25	25	20	20	25	25
Sugar and jaggery	30	30	40	40	30	30	40	40	30	30	40	40

V—Vegetarian NV – Non Vegetarian

Supplementary Foods for Pre-School Children in Developing Countries

Since majority of pre-school children in developing countries consume inadequate diets and suffer from malnutrition, there is an urgent need to institute supplementary feeding programmes in the developing countries. Such supplements should be based on locally available foods such as legumes, oilseeds and nuts. A daily supplement of about 30g of processed legumes or roasted groundnut or processed protein foods based on blends of cereals, oilseed meals and legumes will help to overcome malnutrition among pre-school children.

Exercise

1. Formulate a balanced diet for a pre-school child (1-3 years) by using locally available food stuffs and check whether it takes care of the protein and energy requirements.

2. Formulate a balanced diet for a pre-school child (4-6 years) by using locally available food stuffs and check whether it takes care of the protein and energy requirements.

5. Balanced Diets for Adolescents

The results of diet and nutrition surveys have shown that majority of school children and adolescents consume inadequate diets and are malnourished. The main contributory causes are (i) Inadequate food production, (2) Poverty and (3) Lack of nutrition education. The mean daily nutritional requirements of school children suggested by ICMR. Expert committee is presented in Table 8. Balanced diets at moderate cost for school children and adolescents are given in Table 16.

In view of the shortage in the supply of milk and other protein rich foods of animal origin in the developing countries, it is suggested to use processed protein foods based on locally available oilseed meals and legumes after fortifying with essential vitamins and minerals to supplement the diets of school children.

Table 16 : Balanced diets (g) at moderate cost for school children and Adolescents

FOOD STUFFS	School Children						Adolescents			
							Boys		Girls	
	7-9 yrs		10-12 yrs		13-15 yrs		16-18 yrs		13-18 yrs	
	V	NV	V	NV	V	NV	V	NV	V	NV
Cereals	220	220	290	290	400	400	420	420	320	320
Pulses	70	60	70	60	70	50	70	50	70	50
Green leafy vegetables	75	75	100	100	100	100	100	100	150	150
Other Vegetables, roots and tubers	50	50	75	75	150	150	175	175	150	150
Fruits	100	100	100	100	100	100	100	100	100	100
Milk	600	400	600	400	600	400	600	400	600	400
Fats and oils	30	30	30	30	30	30	40	40	30	30
Meat, Fish and Eggs	--	60	--	60	--	80	--	80	--	80
Sugar and jaggery	30	30	30	30	30	30	30	30	30	30
Peanut (roasted)	30	24	40	30	40	30	54	30	50	50

V—Vegetarian NV – Non Vegetarian

Exercise

1. Formulate a balanced diet of a boy (10-12 years) with locally available food stuffs and check whether it meets the nutritional requirements.

2. Formulate a balanced diet for an adolescent boy and girl (16-18 years) by using most commonly available food stuffs and check for their adequacy of nutritional requirements.

6. Balanced Diets For Adults

Energy

The energy requirements for an adult reference man (60kg) and reference woman (50kg) as per the type of activity are presented in table 17.

Table 17 : Energy Requirement of reference Indian Man and Woman (kcals/day)

Sex	Body weight (kg)	Activity Sedentary	Moderate	Heavy
Man	60	2425	2875	3800
Woman	50	1875	2225	2925

Pregnancy

During pregnancy additional energy is needed to support the growth of the foetus, placenta and maternal tissues as well as to meet the increased Basal Metabolic Rate (BMR). The ICMR expert committee has recommended an additional intake of 300 kcal/day during the second and third trimester of pregnancy.

Lactation

Additional energy required during lactation is computed from the volume of milk secreted, its energy content and the efficiency of conversion of food energy into milk energy. The ICMR expert committee has recommended an additional energy allowance of 550 kcal/day for the first 6 months of lactation. Beyond 6 months, with reduced milk output, upto one year, an extra allowance of 400 kcal/day was recommended for periods from 6 months to 1 year.

Protein

Dietary proteins provide amino acids for the synthesis of body proteins and other biological important nitrogenous compounds in the body. Adequate protein is essential during growth when new tissue proteins are being synthesized. In the adult, dietary protein is essential to synthesize new proteins that will replace those which are constantly being broken down. During pregnancy and lactation, additional proteins are required for the synthesis of foetal tissue proteins and milk proteins, respectively. Dietary proteins should supply the eight essential amino acids (EAA) in the proper proportions needed by the body. The other 12 amino acids present are not considered essential since the body can synthesize them from other carbon and nitrogen sources. The safe level of intake for an Indian adult will be 625 mg of high quality protein per kg when adjusted for the lower quality of dietary proteins with 65 NPU, the safe level of intake in terms of dietary proteins will be 1 g/kg/day.

Pregnancy and Lactation

Protein requirement during pregnancy has been computed on the basis of 'N' accretion for foetal growth and expansion of maternal tissue. The safe intake during the later half of pregnancy recommended by the committee is 15 g/day.

During lactation, protein requirement has been computed on the basis of secretion in milk of 9.4g protein per day during 0-6 months and 6.6 g during 6-12 months, which corresponds to 820 ml and 600 ml of milk, respectively with protein content of 1.15 g/100ml. The committee has recommended during lactation, an additional daily intake of 25g during the first six months and 18g during 6-12 months of lactation. Protein requirement during pregnancy and lactation are shown in Table 18.

Table 18 : Protein requirement during pregnancy and lactation

Group	Daily additional requirement (g)	Total daily requirement (g)
Pregnant women	15	65
Lactating women		
0-6 months	25	75
6-12 months	18	68

Exercise

1. Formulate a balanced diet for a man who is doing heavy work with commonly available food stuffs.
2. Formulate a balanced diet for a man who is doing moderate work with locally available food stuffs.
3. Formulate a balanced diet for a heavy working woman with local food stuffs.
4. Formulate a balanced diet for a pregnant woman doing moderate work.
5. Formulate a balanced diet for a lactating (0-6 months) woman doing sedentary work.
6. Formulate a balanced diet for a man doing sedentary work with economically available foodstuffs.

7. Feeding of Sick Persons

The following factors have to be considered while feeding the sick persons (i) Physiological stress of illness (ii) Nutritional stress, (iii) Effect of illness on food acceptance and, (iv) Acceptability of modified therapeutic diets.

1. The psychological Stress of Illness

The sick person has many fears. These include (i) Outcome of the illness, (ii) Economic concerns for himself and family, (iii) Worry about dependence on others (iv) Restrictions in physical movement and diet during stay in hospital and (v) Monotony due to confinement to bed.

2. Nutritional Stress

Nutritional stresses are due to immobilization and illness. Studies have shown that long term immobilization causes loss of body nitrogen and calcium. The losses of nutrients are also caused by the disease, eg. increased loss of nitrogen and electrolytes in fever, burns, bone fractures etc.

3. The Effect of Illness on Food Acceptance

Many diseases have a profound effect on appetite and food acceptance, illness in general, causes anorexia, nausea and dislike of even foods commonly eaten by the patient. The quantity of food consumed will be small. This will involve frequent feeding with small amounts of the diet.

4. The Acceptability of Modified Therapeutic Diet

The therapeutic diet should be acceptable to the patient and should include the food items commonly consumed by the patient. A real problem will arise if wheat has to be completely eliminated from the diet of the patient (whose staple food is wheat) suffering from malabsorption syndrome as gluten present in wheat aggravates the clinical condition of the patient. In such cases, the patient should be asked to choose rice and corn as substitutes for wheat. Diabetics will have to forego a part of the cereal consumed and a greater part of sugar consumed and eats other foods in its place. Such changes will bring about resistance from the patients. They should be advised that in the interest of their health, they have to accept the modified diets. The patient should be asked to suggest recipes from the food items allowed so that the prepared food items will be acceptable to him.

Other Factors Involved in Feeding the Patient

1. Psychological, social and religious attitude of the patient
2. Congenial atmosphere during feeding.

1. Psychological, Social and Religious Attitudes

The data gathered regarding the attitude of the patient will help to avoid these problems. The doctor, the dietitician and the nurse should do their utmost to make the patient happy.

2. Congenial Atmosphere for Feeding

The time during which meal is served should be such that the atmosphere is quiet and all patients are waiting for their food. The guardian or the attendant staying with the patient can help the patient to eat the diet supplied. If the patient finds any items not to his liking, he or she should inform the nurse and request for an alternate item to be served from the next day.

8. Preparation of Modified Diets Under Selected Conditions for Human

Diets will have to be modified in different ways to suit the needs of patients depending on the nature of illness (i) modifications in diet consistency. (2) modifications in certain nutrient contents and (3) special feeding methods

1. Modifications in Diet Consistency

Diets are modified in three different ways with regard to consistency as follows:

a) Clear fluid diet b) Full fluid diet and c) Soft diet.

2. Modifications in Nutrients

The diets will have to be modified to suit the needs of the patient. The modified therapeutic diet will be prescribed by the doctor.

3. Special Feeding Methods

Patients suffering from severe illness are unable to consume foods and fluids orally. They have to be nourished by special feeding methods. These include (1) Intravenous feeding with glucose, amino acids, vitamins and minerals (2) Tube feeding (Nasogastric feeding).

Modifications in Diet Consistency

Diets will have to be modified with regard to consistency to suit to the needs of the patient. These include: (1) clear fluid diets (2) full fluid diets and (3) soft diets.

1. Clear Fluid Diet

Clear fluid diets are prescribed for patients with marked intolerance to foods as manifested by nausea, vomiting, loss of appetite etc. These consist of tea with lemon and sugar, fat free meat broth, carbonated beverages, fruit juices, glucose water, cereal water and plain gelatin. The clear fluid diets are usually used for 1 or 2 days, till the patient is able to retain and digest a more liberal liquid diet. At the beginning the amount of fluid given to the patient is restricted to about 40 to 80ml per hour, the quantity gradually being increased to 100 to 120 ml as the patient improves. Clear fluid diets are means to provide some calories and make up the water loss in the body. The composition of clear fluid diet is given in the following table.

Table 19 : Composition of clear-fluid diet

Food stuffs	Quantity (ml)
a. Glucose water (10%)	1000
b. Orange juice	1000
c. Barley, water (with glucose)	1000
d. Tender coconut water	1000

2. Full Fluid Diets

Full fluid foods are prescribed when the patient is unable to chew or swallow solid foods. These consist of milk, eggs, cereal porridge, soups, fruit juices and gelatin dessert. This diet provides minimum requirements of proteins, calories, vitamins and minerals. The composition of full-fluid diets are given in the following table:

Table 20 : Composition of full-fluid diets (g/caput/day)

Food stuffs	Vegetarian	Non-vegetarian
Corn flour	50	50
Dhal flour	30	--
Milk	1000ml	800ml
Meat	--	50
Egg	--	one
Fruit Juice	800ml	800ml
Butter	2 table spoons	2 table spoons
Sugar	100	100

3. Soft Diet

The soft diet is used in acute infection, gastro-intestinal disorder and in post-operative cases. The diet is soft in consistency, easily digested and contains very little fibre, spices and condiments. The composition of soft diets is given in the following table.

Table 21 : Composition of soft diets (g/caput/day)

Food stuffs	Vegetarian	Non-vegetarian
Milled cereals and Cereal products (Milled rice, bread, biscuits etc)	300	300
Dhal	50	30
Milk	1000	600
Meat and Fish sausages	--	100
Egg	--	One
Cheese (soft and cheddar)	80	30
Tender vegetables	50	50
Potato	100	100
Tender leafy vegetables	100	100
Fruit (Banana, Apples, Oranges)	100	100
Fats and oils (Butter and Vegetable oil)	30	30
Sugar	80	80

Exercise

1. Suggest a daily menu based on 'Full Fluid Diet' for a patient who is unable to chew or swallow solid foods.
2. Suggest a daily menu based on 'Soft Diet' for a patient who is suffering from gastro-intestinal disorder.

9. Intravenous and Tube Feeding of Human

Patients suffering from severe illness are unable to consume foods and fluids orally. Hence, special feeding methods should be used for feeding such patients. These include (i) Intravenous feeding (ii) Tube feeding (naso-gastric feeding) (iii) Gastroctomy (tube feeding directly into the stomach) and (iv) jejunostomy (tube feeding directly into the jejunum).

1. Intravenous Feeding

The main objectives are:

(a) To provide water and electrolytes to prevent dehydration and correct electrolyte imbalance.

(b) To makeup the loss of tissue proteins and

(c) To provide energy to meet the daily needs of the patients.

Intravenous feeding will have to be resorted to, in the following conditions: (i) In surgery of the gastro-intestinal tract, extensive burns etc. (ii) in cancer of the mouth, pharynx and oesophagus which obstructs the passage of food and (iii) in patients who are unconscious due to hepatic failure, diabetic coma, acute uraemia, injury of the brain etc. The nutrient solutions used in the intravenous feeding are given below.

Table 22 : Solutions for potential nutrition for 24 hrs.

S.No.	Solutions	No. of bottles	Volume (ml)	Amino acids (g)	Calories (kcal)
1.	Fructose (20%)	2	1000	--	750
2.	Amino acids (3.3%) Fructose (15%) and ethanol (2.5%)	5	2500	80	2125
	Total	5	3500	80	2875

The nutrients used in Intravenous feeding are : (i) water and electrolytes, (ii) carbohydrates and alcohol, (iii) amino acids, (iv) whole blood or plasma, (v) emulsified fat and (vi) vitamins.

Water and Electrolytes

The daily minimum water requirement of an adult is about 1.5 litres. Additional water will be required to correct dehydration that might have occurred before the start of therapy. The water loss should be made up in about 24 hours and the urinary excretion restored to about 300 to 400ml in every 8 hrs.

Carbohydrates and Alcohol

The carbohydrates used in intravenous feeding are glucose, fructose and sorbitol. Alcohol can also be given at the rate of 8 g/h as a 3 per cent solution. Alcohol provides 7.1 kcal energy/g. A higher concentration may produce venous thrombosis and intoxication.

Glucose is usually administered as a 5 per cent solution. Fructose is preferable to glucose, as it can be administered at a concentration of 10 to 15% without any adverse effect. Both fructose and sorbitol (30% solution) are rapidly metabolized and their metabolism is less dependent on insulin.

Amino Acids

Amino acid solutions available for intravenous feeding are (i) protein hydrolysates, (ii) synthetic amino acid mixtures and (iii) a mixture of L-amino acids. Out of the above, protein hydrolysates prepared from casein and free from peptides and fortified with tryptophan are commonly used for intravenous therapy.

Whole Blood and Plasma

One litre of whole blood provides about 180g proteins including haemoglobin. Blood is used mainly to combat circulatory failure and loss of blood from body. One litre of plasma contains about 60 to 70g of proteins. Plasma infusion is given as a measure for increasing plasma protein level when it is low.

Fats

Due to various side effects, fat emulsions for intravenous use have been banned.

Vitamins

All the vitamins (twice the daily requirements) should be added to the intravenous fluid to meet the daily needs of the patient.

2. Tube Feeding (Nasogastric Feeding)

The soft polythene tubes of 3-4mm diameter are passed through nose, oesophagus into the stomach after lubricating with oil/ or liquid paraffin. It is essential to make sure that the tube has passed into the oesophagus and not into the trachea. The tube is strapped to the fore head and connected to a gravity drip apparatus fixed in position to a stand; about 2 to 3 feet above the level of the patient and the tube feeds of the following composition are fed (Table 23).

3. Gastrostomy

It may be necessary to feed a patient with carcinoma of the oesophagus through gastrostomy. The tube is inserted directly into the stomach at the time of operation and the above mentioned tube feeds are administered to the patients.

4. Jejunostomy

Feeding through a tube introduced directly into the jejunum will be necessary in case of patients with extensive inflammation and malignancy of the oesophagus and stomach or gasteorectomy. Since salivary and gastric digestion are completely eliminated and the stomach which serves as a reservoir of food is absent. The patients should be given small quantities of feed (100ml) every 30 minutes by using homogenized milk to which glucose and vitamins have been added.

Table 23 : Composition of Tube feeds (g/caput/day)

1. High carbohydrate (no protein) feed

Water	-	1.5 litres
Glucose	-	300 g
Sucrose	-	100g
Orange juice	-	1 litre
Electrolytes	-	as required
Vitamins	-	as required
Total		**2.5 litres**

2. High carbohydrate-moderate fat and protein feed

Milk-cow	-	1.5 litres
Barley water	-	1.0 litre
Glucose	-	300 g
Sucrose	-	100 g
Total		**2.5 litres**

3. High calories-high protein feed

Milk-cow	-	1.5 litres
Evaporated milk	-	1 litre
Glucose	-	200 g
Cane sugar	-	200 g
Total		**2.5 litres**

Exercise

1. Identify various intravenous fluids that are commercially available and write the composition and dosage.

10. Principles and Planning of Balanced Diet Charts for Dog and Cat

Dog and cat-lovers own and rear them as their companions. The importance of the companion animal's role as a component of society continues to gain recognition.

Feeding of Dogs

Dogs belong to order Carnivore, but domesticated dogs have been well adapted to omnivorous feeding habits. Dogs in India are often reared on vegetarian foods without any health problem, provided milk and milk products are included in their diet. The common food stuffs used for the computation of balanced and palatable diets may be classified as follows:

Feed of Animal Origin

1. *Meat:* Variety meats, bone scrapings, meat-cum-bone, edible organs (like heart and kidneys), cleaned and chopped/minced gut etc., meat meal. Meat-cum-bone meal.

2. Milk, milk products and by-products (whole milk, skimmed milk, dried milk powder, casein, cheese meal, paneer, curd etc.)
3. Fish, marine by products and fish meal.
4. Poultry and poultry products/chicken processing by-products (head, legs, gizzard and gut, egg etc.)
5. Edible fatty tissues and fats.

Foods of Plant Origin

1. *Cereals:* Wheat, barley, oat, maize, sorghum, rice, pearl-millet, small millets etc.
2. *Cereal by-products:* Wheat germ-meal, maize germ meal, maize gluten, broken rice, wheat middling, wheat bran etc.
3. *Pulses and byproducts:* Gram, pea, cowpea, green gram, black gram, beans, soybean and their husk removed grits etc.
4. *Fruits and vegetables:* Apple pomace, tomato and sweet citrus fruits, carrot, bean pods, leafy vegetable leaf meals, banana, mango, peaches etc.

Feed Supplements and Additives

Miscellaneous feed supplements are common salt, mineral mixture and vitamin supplements. Main feed additives are antioxidants, bacteriostats, colouring agents (permitted edible colours) and flavouring agents.

Characteristics of Standard Dog Foods

The diet should be balanced for essential nutrients and should be formulated according to the requirement of dogs viz.; a low energy diet is required for idle adult dogs, high energy diets for working dogs and high energy-high protein diets for growing pups. It should be cooked except in cases where other treatments like pasteurization, irradiation etc., have been suggested. Green vegetables should be half boiled.

Precautions in the Feeding of Dogs

* Pup's diet should be liquid or semi-solid.
* Raw meat, bone, fish and chicken may transmit many infections and should not be fed without cooking. Raw eggs contain antivitamin 'Avidin' and hence should be boiled and fed.
* Drinking water should be offered fresh and changed frequently.

* A definite feeding schedule should be followed and it may be divided as light breakfast, lunch and dinner.
* Diet should be prepared from a variety of feeds but there should not be frequent changes in the diets and feeding schedule.
* Dogs should not be allowed to develop liking for a particular food because it may cause malnutrition leading to either obesity or undesired leanness.

Table 1 : Daily Dry-matter requirement for maintenance (on Dry-matter Basis)

Body weight (kg)	Dry matter requirement (g/kg body weight) 3.3 kcal/g	3.5 kcal/g	4.0 kcal/g
2.5	33	31	27
5.0	28	26	23
7.5	25	23	21
10.0	23	22	19
15.0	21	20	17
20.0	18	17	15
25.0	18	17	15
30.0	17	16	14
40.0	17	16	14
50.0	15	14	13
60.0	15	14	13
70.0 and above	14	13	12

• Calculated from the values of NRC (1974) for diet containing 3.3 kcal ME/g dry matter.

Long fresh bones with open extremities should be offered to chew for half to one hour to keep teeth and jaw muscles healthy. Small bones and splinters in meat and fish pieces should be avoided in dog meals. Spoiled foods and damaged grains should never be fed because they are potential sources of infections and toxicosis.

Meat obtained from hormone-implanted animals and chemically caponized birds should not be used for feeding of breeding and growing dogs.

Food Requirement and Balanced Diet Charts

A standard balanced dog food should contain 20-22% CP, 4-5% fat, 5-10% CF and 4-5% ash on DM basis. Average energy content should be 3.3-4.0 Mcal per kg of food on DM basis. It should be balanced for essential minerals and vitamins. The DM requirement of dogs for maintenance is given in Table 1.

The requirement for production i.e., growth, work, pregnancy and lactation are higher than for maintenance. Diets supplying 3.5 to 4.0

kcal ME per gram dry matter are fed for production. Young pups in early life may require even higher energy content in their diets for optimum growth and development. Average ME requirement of mature idle dog is 132 kcal per kg $W^{0.75}$ which increases to 188 and 470 kcal/kg $W^{0.75}$ in terminal trimester of pregnancy and lactation period, respectively. The average requirement of growing and adolescent puppies is 274 and 200 kcal/kg $W^{0.75}$ respectively.

Feeding of Pups

Under normal conditions new born pups are nursed by their mother through natural teat feeding and they learn to eat mother's food after opening eyes at about 2 weeks of age. Lactation period may last for more than 8 weeks but pups are usually weaned at about 4, 6 or 8 weeks of age. Weaned pups are gradually shifted to artificial feeding of milk-based semi-solid or liquid diets. Artificial feeding is required for orphan pups as well as those pups whose mothers are unable to nurse them.

If more than 8 pups are born at one time, the litter should be reduced to 8 at the end of 1st week. With large litters it is necessary to supplement the bitch's milk with cow's milk after 3rd week. For a litter of 8 pups, a pint of milk is sufficient daily for the 1st 3 days, after which they receive a quart. At 5 to 6 weeks of age, the pups can be weaned and can receive cow's milk and solid food. At 8 weeks of age the pups can receive the same food as mature dog. Orphaned pups can be fed as follows:

Cows milk	=	800ml
Cream (12% fat)	=	200ml
Egg yolk	=	1
Egg white	=	6g
Vitamin A	=	3000 IU
Vitamin D	=	500 IU
Citric acid	=	4g and
Steamed bone meal	=	Sufficient quantities

This is fed 6 times/day for 3 weeks followed by 5 times a day till they start eating solid food.

Milk Yield and Composition

Milk yield in bitches is significantly influenced by the breed, age at 1st breeding, body condition, litter size and stage of lactation.

Average daily milk in small, medium and large sized breed is about 2.4, 2.7 and 2.8 per cent of their body weight in the beginning with an increase up to 3.4, 4.6 and 3.6 per cent of body weight at 4 weeks of lactation period; followed by gradual decline by weaning of pups of 6-8 weeks of age.

Table 2 : Feed formulae for young pups

Ingredients	Diet 1	Diet 2	Diet 3
Dried milk powder	200	--	--
Liquid whole milk			
Cow	--	700	--
Buffalo	--	--	800
Cream (30% fat)	200	200	100
Egg-yolk	--	50	50
Boiling water	400	--	--
Rice gruel/flour gruel (30% dry matter)	200	50	50

- For every 1 kg of liquid diet containing about 25-30% dry matter add about 2000 IU vitamin a, 500 IU vitamin D and 4g citric acid.
- Antibiotics preferably tetracycline @ 10mg per kg feed should be mixed.

Table 3 : Feeding schedule of growing dogs

Body weight (kg)	Amount of various feeds (g/day)			
	Cereals	Meat	Milk	Fresh bones (steamed)
5	200	100	100	One metacarpal/ metatarsal/ humerus/femur of goat or sheep
10	400	200	100	
15	600	300	100	
20	800	400	100	
25	1000	400	100	

- About 50-100 g green leafy vegetable (half boiled) or fresh fruit and an egg should be fed daily in the absence of meat/milk.

Table 4 : Feeding schedule of mature dogs

Body weight (kg)	Daily food requirement (g)			
	Cereals	Meat	Green Vegetables of legume pods	Milk
5	100	50	50-60	100
10	150	50	60-80	100
15	200	100	80-100	100
20	300	100	100-150	--
25	400	100	100-150	--
30	450	150	100-150	--
40	500	150	150-200	--
50	600	200	200-300	--
60	700	200	200-300	--
70 and above	800	200	200-300	--

- 1-2 eggs should be fed daily if meat/milk is not given.

Feeding of Pregnant Bitch (in third trimester)

Since active development of litter takes place in the terminal trimester of the gestation period, the pregnant bitches should be given about 20-39 per cent more food than maintenance requirement after 6 weeks of gestation period. The ration schedule is presented in the following table. Diets containing more than 1600 kcal of ME per kg of food with 21% are recommended. To ensure proper nutrition one should provide good quality labeled dog food at all stages of life.

Table 5 : Feeding schedule of pregnant bitch

Body weight (kg)	Daily food requirement (g)				
	Cereals	Meat	Milk	Green vegetables	Bone (Steamed)
10	200	50	100	100	Fresh long bone of sheep, goat, cattle or buffalo
15	250	100	100	100	
20	350	100	100	100	
25	450	150	100	150	
30	550	150	100	200	
40	600	150	--	200	
50	700	200	--	250	
60 and above	800	200	--	250	

Feeding During Lactation

The demand for milk by nursing pups continue to increase for about 20-30 days. At peak lactation, the food intake may be 2-4 times above the normal maintenance intake. The diet used during gestation period may be fed during lactation. Pups are weaned at 6-8 weeks of age and by this time the female's food intake should be less than 50% of her usual or maintenance level.

Feeding of Working Dogs

The energy requirement of working dogs engaged in strenuous work like pulling of sledge and running more than 100 km for hunting wild game may need as high as 3-4 times of their maintenance requirement (Table-6). The mean ME content of diet should be 4 Kcal per g of food on dry basis.

Table 6 : Feeding schedule of Working dogs

Body weight (kg)	Daily food requirement (g)			
	Cereals	Meat	Vegetables	Bone (steamed)
30	1400	500	300	One long bone of fresh carcass of cattle or buffalo
40	2000	500	300	
50	2500	600	300	
60	2800	600	400	
70	3000	600	400	

- The amount of cereals is partly replaced by adding edible fat @ 1g for each 2.5g cereals and percentage of protein is increased accordingly.

Feeding of Stud Male

Male dogs reared exclusively for breeding require more critical diets than the working dogs. Stud males are fed according to the feeding schedule given for maintenance except during the breeding season. Feed level should be increased by 25 per cent of the maintenance requirement by 3-4 weeks before the mating or semen collection. Dogs should also be given a boosting dose of vitamins A, E and B-complex either orally or in feed.

Feeding of Fat (Obese) Dogs

To reduce the energy content, the diet of obese dogs should be prepared from high fibre-low energy feeds. A feeding schedule given in Table 7 may be used with minor modifications depending on the degree of fatness of the dog.

Table 7 : Feeding schedule for obese dogs

Body weight (kg)	Daily food requirement (g or kg)			
	Lean meat	Skim milk	Green beans/vegetables	Bran-rich bread
20	100	--	200	50
30	150	--	200	50
40	150	100	300	100
50	200	100	300	150
60	200	200	400	200
70	200	200	500	300

- Oral mineral vitamin supplement should be administered daily.

Table 8 : Dog food categories

Growth foods-Growth or All Life Stages Claim (Appropriate for reproduction and hardworking dogs)	
Protein	21% or greater
Fat	8% fat or greater
Adult Foods All Life Stages Claim or Maintenance Claim	
Protein	21-26%
Fat	8-18%
Reduced Calorie Foods Maintenance Claim	
Protein	14-25%
Fat	7-8%
Performance Foods All Life Stages Claim	
Protein	27% or greater
Fat	9% or greater

Exercise

1. What is the dry matter requirement of the following class of dogs with examples of economic feeding schedule

   a. Growing dog of 10kg body weight

   b. Mature dog of 20kg body weight

   c. Pregnant bitch of 25kg body weight

   d. Obese dog of 40kg body weight

   e. Suggest an economic feed formula for feeding young pups.

Feeding of Cats

Cats under normal habitat are opportunist feeders eating as and when they catch and kill prey. Hence, they are adapted to large meals at irregular intervals. However, it is likely that small rodents and birds which would constitute the major part of their diet are not big enough for one kill to provide for the cat's daily needs and more than one meal per day would be more useful.

Domestic cats on palatable foods prefer to eat many small meals (12-20 meals per 24 hours) rather than one or two large ones. Most cats regulate intake so that they do not overeat and become obese when food is made available in this way. Many pet cats are fed 'on demand' with food made available whenever the animal asks for it throughout the day. The average pet can weigh about 4 kg and needs about 350 kcal metabolizable energy per day. This would be provided in 90g dry matter (100 g dry food).

Kittens at birth usually weigh between 80 and 140g with an average weight of about 100g. They are entirely dependant on the milk of queen for about 4 weeks by which time they treble their birth weight. From this age onwards, the process of weaning or gradual replacement of queen's milk by other foods can begin. By fifth week they begin to eat the queen's food and will start to eat finely minced or chopped moist food. By the time the kittens are 7-8 weeks old the proportion of their total nutrient intake coming from supplementary food should be at least 70-80% and they can be finally separated from their mother and fed independently there after.

A weanling kitten at 7-8 weeks of age weighs anything from 600g to 1 kg. The energy requirements of growing kittens are available depending on size, activity and environment but not less than 200 kcal ME per kg live weight. Kittens grow very rapidly if fed generously and will achieve adult weight of around 3.5kg by 6 months of age. At this stage they are still very active and will usually eat enough food to supply 120-150 kcal/kg and by one year of age they settle down to an average adult intake of 85-90 kcal/kg. Kittens very quickly reach intakes of 300-400 kcal per day and grow rapidly without increasing their daily intake to any degree. Intake is usually maintained at this level until maturity although growth gradually ceases and cats achieve a mature weight between 8 and 12 months of age. It is good to introduce different varieties of flavours while they are still growing so that they will readily accept a variety of foods as adults.

Feeding cats in Pregnancy and Lactation

There is no need to provide any special feeding for cats which are pregnant, if a regime of feeding to appetite with a variety of foods is being followed. The extra nutritional needs of pregnancy are small and will be adequately catered by normal prepared foods.

Once the kittens, have been born the nutritional needs of the queen increase rapidly because she has to provide through her milk the nutrients and energy needed for very rapid growth of her offspring until they begin to eat supplementary foods when about 5 weeks old. Depending on the number of kittens, this demand reaches 3 to 4 times her normal maintenance needs. The queen should be encouraged to increase her nutrient intake by the provision of frequent meals and by offering more concentrated foods. It is good practice to offer milk as well as water at least twice a day, to provide free access to dry or semi-moist food and to give four or more meals of canned food daily.

Special Feeding Characteristics of Cats

The peculiar metabolic functions of cat suggest that cat has been developed as a special carnivore. Other carnivores like dogs do not exhibit similar nutritional needs. The differences are discussed below:

1. Essential Fatty Acid (EFA) Nutrition

In most of the mammals the EFA linoleic acid which is found in abundance in vegetable oils can be converted to archidonic acid. However, the cats have only a limited ability to convert the parent EFA into the longer chain derivative (arachidonic acid). Hence, cats require a dietary source of preformed 20:3 or 20:4 acids which means certain need for EFA of animal origin. NRC recommendation for linoleic and arachidonic acids for cat are 1% and 0.1% of dry matter (approximately 2.3% and 0.23% of calories), respectively. The recommendation for the dog is 1% linoleic acid only.

2. Amino Acids

The adult cat is different to dog in that it requires a dietary source of the amino acid arginine. Arginine deficiency in cat rapidly results in severe adverse effects because of the inability to metabolic nitrogen compounds (via the urea cycle), which gets accumulated in blood stream as ammonia (hyperammonaemia) and in serious cases, can lead to death within hours. The rapidity of the effects is second only to a lack of oxygen. This unique requirement appears to be due to an inability to synthesise the amino acid ornithine (a component of urea cycle) since the latter protects cats against the adverse effects of arginine deficiency.

Table 1 : Recommended amino acid levels for dogs and cats

Amino acid	Dog		CAT
	Adult	Growing	Growing
Theronine	264	600	680
Valine	325	450	510
Cystine(a)	88	350	380
Methionine	180	450	380
Cystine+methionine(a)	268	800	760
Isoleucine	302	800	510
Leucine	544	600	1020
Tyrosine (a)	211	--	425
Phenylalanine	304	--	425
Tyrosine+phenylalanine(a)	515	1000	850
Histidine	132	130	255
Tryptophan	79	150	125
Lysine	638	650	680
Arginine	NR(b)	1000	900

All values are expressed as mg per 400 kcal ME, which approximates to mg/100g dietary dry matter in commercial foods.

- The values shown for cystine and tyrosine reflect the extent to which these amino acids can substitute for the total requirement for methionine and phenylalanine, respectively.
- NR= not required. Arginine is, however, required by the adult cat.

3. Taurine Synthesis

Taurine is useful in the metabolism of sulphur containing amino acids. Cats have little capacity to synthesize taurine from cysteine. Taurine is found in all tissues and is particularly high in retina and lens. When there is no dietary supply of taurine, cats suffer from central retinal degeneration and subsequent blindness. In cats some of the cysteine flux is directed towards the synthesis of felinine and isovalthine (B and C carbon addition products of valine and cysteine) peculiar to felines.

Taurine is not an amino acid but an amino sulphonic acid which is not a part of the polypeptide chain of protein. It is an end product of sulphur amino acid metabolism and is produced from sulphur containing amino acids methionine and cystine. Unlike other mammals cats cannot synthesize enough taurine from sulphur containing amino acids. The enzyme responsible for the conversion is not absent totally in cats but its activity is not high enough to supply all of the cats needs. Cats are also totally dependent on taurine for the formation of bile salts; unlike other species it cannot also use glyine for 0.05% of dry matter, i.e. about 50mg per 400 kcal. The richest sources of taurine are animal derived raw material and little is found in plant tissues. This is another important nutritional difference between dogs and cats.

4. Conversion of β carotene to vitamin A

The vitamin A needs of most of the mammals including human can be met from foods of plant origin because of their ability to convert β-carotene to vitamin A. Cats lack this capacity and hence need preformed vitamin A which occurs only in the foods of animal origin.

5. Synthesis of niacin from tryptophan

The vitamin niacin can be synthesized from the essential amino acid tryptophan by most of the mammals. Feline species however are unable to synthesize nicotinic acid from tryptophan despite the presence of all the enzymes of the pathway. The reason appears to be very high activity of the enzyme picolinate carboxylase which channelizes tryptophan towards glutamic acid synthesis (Non-niacin pathway) rather than nicotinic acid biosynthesis.

Vitamin C

Unlike man, dog and cats do not need a dietary supply of this vitamin because they are able to synthesize it from glucose. Hence, there is no need to include this nutrient in foods for dogs and cats.

Ration for cats

1. Dried milk powder	=	100g
2. Skimmed milk powder	=	300g
3. Cream (30% fat)	=	50g
4. Rice gruel	=	150g
5. Boiling water	=	400g
Total	=	1000g

Table 2: Cat food categories

Growth foods	–	Growth or All Life Stages Claim
Protein		30% or greater
Fat		8% fat or greater
Adult Foods	–	All Life Stages Claim or Maintenance Claim
Protein		30%
Fat		6%
Reduced Calorie Foods	–	Maintenance Claim
Protein		30%
Fat		6-8%

11. Types of Pet Foods

Selection of food for pets depends on several factors like economic status of the owner, nutritional requirement of pets for different physiological states of the body and the feeding value of different foods. A diet may have one of the following 3 types of food or their combinations.

1. Kitchen Residue

Usually all types of food normally eaten by man are also eaten by dogs. This contains a combination of all cooked food. Indian vegetarian diets are normally rich in carbohydrates and fat, but poor in good quality protein foods. Such kitchen waste should be adequately supplemented with milk and milk products, eggs, meat and fish to balance the diets.

2. Home made foods

Most of the home made diets are unbalanced especially for protein and minerals due to the use of larger proportions of cereals and starchy tubers and corn in the human food. Freshly prepared home made diets are generally more palatable than the proprietary foods. A balanced home-made diet should contain about 20-22 per cent protein (at least one third from milk, meat or egg), 5-10 per cent fat, 5-10 per cent fibre and 3.5-4.0 Kcal ME per gram of diet on dry matter basis. In addition, adequate minerals and vitamins should be fed. These foods are prepared by baking, roasting, boiling and cooking. The following are the two model rations.

Ingredient	Ration I (%)	Ration II (%)
Yellow corn	35	58
Wheat shorts	20	20
Wheat bran	10	–
Fish meal	10	–
Meat scrap	10	20
Skim milk	10	–
Alfalfa meal	2	–
Bone meal	2	–
Salt	1	1
Cod liver oil	–	1
Total	100	100

3. Commercial pet foods

Owners usually prefer to feed their pets a standard commercial diet which is balanced and prepared specially for pets. Generally, 3 types of commercial pet foods are marketed. They are (1) Dry foods, (2) Soft-Moist foods and (3) Canned products. These vary in a number of characteristics including moisture, cost, palatability and amount of nutrition delivered per pound of food. Inspite of these variations, all types of pet food products can be formulated to provide complete and balanced nutrition for dogs and cats.

i. *Dry Diets*: Most dry dog foods contain 18 to 27% protein, 7 to 15% fat, less than 12% moisture and 35 to 50% carbohydrate (NFE). These diets deliver between 1400 and 2000 metabolizable kilocalories per pound of diet. They are made from cereals and cereal by-products, protein concentrates of animal or vegetable origin. Meat and bone meal, fish meal, soybean meal, fats, mineral and vitamin supplements.

Composition of Dog Biscuit

S.No.	Ingredient	Amount (parts)
1.	Ground corn meal	43
2.	Wheat grain	20
3.	Meat scraps	10
4.	Skim milk powder	10
5.	Soybean meal	5
6.	Groundnut meal	5
7.	Alfalfa meal	2
8.	Bone meal	2
9.	Iodized salt	2
10.	Brewer's yeast	1
	Total	100

It is generally accepted that dry pet foods are more economical to feed than the soft moist and canned varieties and offer more convenience to pet owners. They contain greater concentration of nutrients and energy per unit weight than foods of high moisture content and hence small amounts are needed to provide particular quantity of nutrients. These are sold as baked biscuits, extruded and expanded biscuits, extruded in solid cubes or cylinders like farm animals feeds or as mixtures of meal and flakes.

ii. *Soft-moist diets* : Soft moist dog and cat foods contain approximately 16 to 25% protein, 5 to 10% fat, about 25 to 35% carbohydrates and 30% water (moisture level sometimes as high as 50%). High quality soft-moist diets contain approximately 1200 to 1350 metabolizable kilo calories per pound of product. Soft moist dog and cat food generally offer higher palatability as compared to dry diets, are convenient to serve and store easily but are more expensive than dried diets. These diets are made with meat, meat by-products, soya or other vegetable proteins, cereals, fat and sugar.

iii. *Canned Diets*: Canned pet food contain between 8 and 15% protein in dog foods and 12 to 20% protein in cat food products. Both types of diets contain between 2 and 15% fat, depending on the type(s) of animal tissue used in the diet. The moisture content for canned pet foods is approximately 75%. Canned dog and cat foods deliver 375 to 950 metabolizable kilo calories per pound. Total food consumption is high because of low calorie content. The nutrient balance is dictated to a great deal by the type of animal tissue used in the diet. These diets offer the highest palatability and the higher cost per serving when compared to the dry and soft-moist products. In addition, canned products are less convenient to serve than other product types. Once a can is opened, unused food should be stored in a refrigerator. The main ingredients of canned food for both species are meat, meat by-products, vegetable proteins, cereals, minerals and vitamin supplements.

Comparison of pet food types

Product characteristics	Dry diets	Soft moist diets	Canned diets
Cost	Low	Moderate high	High
Palatability	Moderate	Moderate high	High
Moisture content	12%	20-25%	75%
Metabolizable Kcal/pound	1400-2000	1200-1350	375-950

Table 1: Expected average daily intake of canned, dry and semi-moist foods by cats of differing weight and status

	Expected daily intake			
	Canned food Av. Quantity			
Weight of cat (kg) Young growing cats	**g/day**	**cans**	**Dry food g/day**	**Semi-moist food g/day**
1	265	2/3	60	70
1.5	400	1	90	100
2	470	1 ¼	110	120
2.5	500	1 ¼	115	125
3	535	1 ¼	120	135
Adult cats				
2.5	300	¾	70	75
3	360	¾+	80	90
3.5	420	1	95	105
4	480	1+	110	120
4.5	540	1 ¼	120	135
5	600	1 ½	135	150
Lactating queen peak lactation	1080	2 ½	245	270
3	1260	3	290	315
3.5	1450	3 ½	330	360
4				

Table 2 : Suggested amounts of various foods needed to meet the average energy needs of growing puppies and young dogs.

Weight of puppy (kg)	Canned meat for puppies plus mixer biscuit 3:2 by volume				Type of food*					
	Mixture	Cans	+	g biscuit	Canned Puppy food only*	Canned food (cans)	+	Biscuit 1:1 By volume (g)	Semi-moist dog food (g)	Complete dry dog food (g)
1	210	½	+	40	?					
2	345	?	+	65	1				180	145
3	470	1	+	85	1 ½				245	195
4	590	1 ¼	+	100	2				305	245
5	690	1 ?	+	125	2 ¼				360	290
6	790	1 ½	+	145	2 ?				410	330
7	880	1 ¾	+	165	3				460	370
8	980	2	+	180	3 ¼				510	415
From 10kg feed 1:1 by volume										
10	1060	2	+	270		2 ¼	+	320	610	490
12	1210	2 ¼	+	300		2 ½	+	365	700	560
14	1360	2 ½	+	350		3	+	410	780	630
16	1500	2 ¾	+	375		3 ¼	+	450	860	695
18	1650	3	+	400		3 ½	+	495	950	760
20	1800	3 ¼	+	450		3 ¾	+	530	1000	820

* Canned puppy food ('Pedigree Chum food', United Kingdom) assumed to be 110-112 kcal/l00g. Biscuit 340/350 kcal/100g
Complete dry food = 380 kcal/100g Semi-moist dog food = 310 kcal/100g

Table 3: Average energy needs of normal adult of dogs and amounts of food needed to meet them
1. Canned food + biscuit

Weight of dog (kg)	Typical breed	Daily energy needed (kcal)	Food to provide this amount of energy Canned meat and biscuit mixtures, equal volumes or 3:1 by weight and 2:1 by weight					
			3:1			2:1		
			Total wt (g)	Cans	Biscuit (g)	Total wt(g)	Cans	Biscuit (g)
2	Yorkshire Terrier	240	155	¼	40			
6	Cairn	540	350	?	85	305	½	100
13	Beagle Cocker	900	585	1	145	505	¾	170
20	Samoyed	1250	810	1 ½	200	702	1	235
25	Collie	1500	960	1 ¾	240	830	1 ¼	280
30	Boxer Labrador	1700	1100	2	275	955	1 ½	320
35	German Shepherd Dog	1900	1235	2 ¼	300	1070	1 ?	360
40	German Shepherd Dog	2100	1365	2 ½	340	1180	3 ¼	400
50+	Great Dane, Newfoundland	3500	2270	4	570	2000		600

Note

- Energy needs increase by about 45 kcal for each kg increase in body weight from 20 to 40 kg
- Values are average estimates based on moderate exercise and reasonably warm housing. Individual dogs may need 25% more or less than the amounts suggested.
- For large breeds with good appetites, the proportions of meat and biscuit may be altered from equal volume to 1 part meat to 2 parts biscuit, i.e. 3:1 to 2:1 by weight.

Table 4: Average energy needs of normal adult dogs and amounts of food needed to meet them.

2. Semi-moist & dry foods

Weight of dog (kg)	Typical breed	Daily energy need (kcal)	Semi-moist food needed to provide this amount when fed as the only food (g)	Complete dry food needed (g)
2	Yorkshire Terrier	240	75	60
6	Cairn	540	170	140
13	Beagle Cocker	900	280	230
20	Samoyed	1250	390	320
25	Collie	1500	460	380
30	Boxer Labrador	1700	530	440
35	German Shepherd Dog	1900	590	490
40	German Shepherd Dog	2100	660	540
50+	Great Dane, Newfoundland	3500	1100	900

* Giant breeds appear to need almost 200 kcal/$W_{kg}^{0.75}$ per day for maintenance or 50% more than dogs in the medium-sized range.

12. Feeding of Sick Dogs and Cats

Precautions have to be taken while selecting foods for pets suffering from systemic and other diseases.

Liver diseases

In case of liver diseases easily metabolizable feeds like dextrose, rice and maize starch should constitute the source of energy and protein of high biological value. They should be feed at the rate of 3.5-4 g per kg body weight. Feeding of fish, fish meal and meat products should be avoided. A multivitamin preparation should be administered daily. In cases of chronic liver diseases and ascitis feeding of high sodium feeds and common salt should be reduced drastically.

Hypoglycemia

In acute hypoglycemia glucose or sucrose should be fed at the rate of 70g per kg body weight once or divided into 2 or 3 parts for feeding at intervals of 2-3 hours. The daily diet should be prepared from maize starch, rice, oat-meal, egg yolk, meat meal and mineral vitamin supplement. Strenuous work should be avoided in working dogs.

Diabetes mellitus

Quality of skimmed milk, green vegetables, fibre (bran) enriched cereal foods should be increased. The proportion of starchy foods and sugars should be drastically reduced from pet's diet.

Diarrhoea

The diet should contain low fibre cereals, low protein and easily digestible protein foods like paneer, cheese, curd, eggs and liver. Proprietary diets containing flesh and high concentration of sugar, lactose and dried whey should be avoided. Eggs can be fed if there is K deficiency and in acute cases potential fluid feeding may be necessary to restore the body fluid volume and electrolytes. The following recipe may be useful in diarrhoea at the rate of 100ml per kg body weight daily in 5-6 split meals at equal intervals.

Food ingredients	**Quantity**	**Dry matter**
Cooked rice gruel or low fibre wheat flour gruel or sago gruel	500ml	100g
Sugar	60g	60g
Paneer or cheese	200g	50g
Cream	100g	30g
Maize oil	15g	15g
Hard cooked egg	100g	50g
Dicalcium phosphate	10g	10g
Potassium chloride	15g	15g
Mineral vitamin supplement	as per manufacturers recommendation	

Gastro enteritis

Irritant diets should not be fed. The diet may be prepared from chicken soup, egg white, milk, sago, vegetable soup, sugar and cream supplemented with dicalcium phosphate, trace minerals and vitamins.

Vomiting

Vomiting may result from eating of spoiled foods or in diseased condition. Normally feeding should be avoided for the day. Dogs should be offered fresh light diet as suggested for diarrhoea.

Ascites

Restriction of sodium intake or withdrawal of sodium from the diet, egg white, solid high protein and fibre diet should be given, liquid diet should be reduced.

Cardiac insufficiency

Salt should be reduced drastically and as far as possible low sodium foods should be used. Lean meat, fresh water fish, egg yolk and rice should be fed without salt. Milk, cream, butter, fish, nuts and eggs should be avoided.

Renal insufficiency

High protein diets should be restricted and minimal amount of good quality protein of high biological value should be fed along with high energy food. The purpose of various dietic measures is given below:

S.No.	Dietetic measures	Intention
1.	Adequate water intake	Optimum hydration
2.	Adequate sodium supply	Compensation of sodium loss via the damaged kidney
3.	Controlled potassium intake	Prophylaxis against hyperkalaemia
4.	Adequate energy supplied in the form of carbohydrates and fat (up to one third of total dietary energy as fat)	Prevention of protein being used as an energy source and thus contributing to uraemia
5.	Control of protein intake	
	a. Restriction where indicated	Reduction of protein breakdown products in tissues (eg. Urea, hydrogen ions)
	b. by providing good quality protein supplements	To compensate for heavy protein losses via the damaged kidney

Urolithiasis

The diet should be prepared on the basis of composition of uroliths detected in urine. Low protein, low mineral food containing high percentage of common salt should be fed to increase water intake and urine volume. An increase in urination, frequency and urine volume reduces the urinary concentration of involved minerals.

Pyrexia (fever)

Energy requirement of pets increases in fever. They should be fed additional energy at the rate of 7 Kcal per kg body weight daily for each degree rise above the normal body temperature. This can be accomplished by feeding cream, hard cooked egg or maize oil in the diet.

Pica (depraved appetite)

Due to phosphorus deficiency, animals develop a habit of seating unusual substances like decaying bone, wooden pieces, grass etc. In such cases the diet should be balanced through mineral supplementation and they should be offered fresh long bone of healthy carcass.

13. Feeding of Laboratory Animals

Nutrient requirements of laboratory animals can be met through feeding balanced diets which provide adequate amounts of all the nutrients such as protein, carbohydrates, fats, vitamins and minerals. Laboratory animal's diets are of 3 types.

1. Natural diets

Comprising of a blend of cereals and legumes supplemented with vegetables and fruits.

2. Semi-synthetic diets

These are made from a combination of natural ingredients and are practically purified (starch, sugar, casein, fat, vitamins and minerals). Most of the diets belong to this category.

3. Synthetic diet

Include combination of purified protein, amino acids, carbohydrates, fats, vitamins and minerals. These are very expensive and are used for special type of experiments.

The selection of ingredients for formulating diets is governed by availability, cost and quality. Frequent analysis of the diet for proximate analysis will ensure its nutritional quality.

Formulation of feeds

A feed should contain ingredients which will provide all the necessary macro and micro nutrients in right proportions. Nutrient requirements of laboratory animals are given either per unit quantitative diet (generally per kg of diet) or as the daily requirement for the animal. It is not desirable to use large proportion of a single ingredient. Diets should contain ingredients available under local conditions incorporating a wide variety of foodstuffs.

The formulation of most of the diets comprise of cereals or purified starch or purified starch or sugars as carbohydrate source. Cereals contain 10-12% protein. Legumes such as groundnut, soybean and other pulses are important sources of vegetables proteins. However to supply all the essential amino acids in the diet in proper proportions it is necessary to add animal protein. Foods such as skimmed milk powder, meat, fish meal, bone meal and casein are good protein sources of animal origin.

Important sources of fat in the animal diets are groundnut oil, corn, oil coconut oil and tallow (animal fat). However, tallow is poor source of linoleic acid. At least 1% of dietary calories. Meal in the diet would ensure the supply of linoleate. Diets formulated using natural ingredients like cereals, pulses and legumes contain some amount of minerals and vitamins and their content can be increased by supplementation with mineral and vitamin mixtures.

Preparation of diets

Laboratory animal diets can be prepared in the form of dry powder, wet mash or pellets. Nutritional adequacy, palatability, stability on storage is essential requirements of good diet. The diets may be mixed with water to form a paste and can be fed in that form. Mashes sour quickly, especially at animal room temperatures and within a day or two will also grow mould. A wet mash, however, can be dried some times to form as a hard cake or biscuits, the water content of which may be 5%.

Now a days the laboratory animals are fed on compound compressed pellets. These pellets are made by mixing the ingredients together injecting steam and forcing the mixture through a suitable diet. The heat generated in the process kills most of the bacteria. Pelleted diets put in the cages of animals are not soiled with the bedding and droppings. Badly made pellets are likely to be friable leading to wastage. The size of the pellets for different species varies. For rats and mice the pellets of 8-10mm diameter are found to be suitable and for guinea pigs 4-6 mm diameter are preferable.

Daily intake of diet and water

The average quantity of balanced diet and water consumed by different species of laboratory animals are given below:

Species	Food (g)	Water (ml)
Rat	12-15	35
Guinea pigs	30	150
Rabbit	150	300
Cat	100-200	320
Dogs	300-500	350

These quantities are normal for adult animals but pregnant, lactating and young animals require higher quantities of diet. Lactating animals consume more water and they should be supplied with *ad libitum* water. Reproducing and lactating animals need extra nutrients and should be supplied either by increasing the ration or by supplementation.

Assessment of quality of feeds

An animal must eat to sustain growth, physical activity, body heat and reproduction. Digestive efficiency differs between animals of the same species and such differences exist at different periods of life. Changes from one diet to another diet should be gradual in a period of 10-15 days.

The theoretical composition of diets as calculated from the food tables may differ substantially from the actual composition of the food eaten by the animal. These differences arise due to variation in the composition between the batches of the ingredients, deterioration during storage etc. Gross imbalances between the bulk ingredients (protein, carbohydrates, fat, indigestive residues) and deficiencies of accessory factors (vitamins and minerals) should be avoided.

Evaluation of feeds

A diet must be evaluated by assessing its quality. Weight gain is measured in animals from weaning till they attain maturity. The assessment of diet by reproductive performance is the most stringent test of dietary sufficiency, but it may be necessary to go through several litters and at least 3 generations to obtain significant results. Chemical analysis of nutritional qualities of diet for major and minor nutrients should be carried out periodically to ensure its quality.

Adequate diets of laboratory animals should meet the requirements during growth, generation and lactation, work, maintenance and other conditions, commercial diets provide the most uniform nutrition. However, some variation may occur due to processing methods and storage conditions.

Types of diets

1. *Pure chemical nutrient diets*: These are mixtures of carbohydrates, amino acids mixed together. They are expensive and generally used under highly controlled conditions and usually for short periods.

2. *Purified diets* : Pure diets consists of mixtures of ingredients including chemically pure salts, vitamins, sucrose, fats and proteins such as casein with known amino acid composition. Purified diets are less expensive than pure chemical diets and are generally used for nutritional studies. These diets are useful since specific nutrients can be decreased in the basic diet and then added back in graded levels and the comparative levels fed simultaneously with the specific nutrient as major variable.

3. *Commercial diets*: These are digested and tested for their ability to provide the precise nutrients needed by specific animal species.

4. *Therapeutic diets* : These are used for stress conditions encountered normally and include for example (i) low sodium diets for cardiac problems (ii) low caloric density diets for obesity, (iii) low protein diets to prevent kidney nitrogen overload etc.

5. Dietary additives: Additives are added to the diets for different purposes. Some enhance nutrient availability, protect against disease, add flavour and aroma and retard oxidation.

Feeding methods

1. *Ad libitum vs* restricted feeding: When the feed intake of the animal is restricted, this process of feeding is called 'restricted feeding while unrestricted feeding is called *ad libitum* feeding'.
2. *Controlled feed intake*: This method of offering feed influences the feed intake. Most species prefer moist feeds than dry feeds. Adding controlled levels of moisture is one way to increase feed intake in domestic and laboratory animals. With most dry diets, addition of one part by weight of water to 2 parts by weight of dog food will increase food consumption by about 20%.

The addition of 20% of diet as meat will increase the total nutrient intake by another 20%. These methods are useful when increase in nutrient intake is required during the stress of lactation and hard physical work. Food intake can also be controlled by offering measured quantities of food.

14. Balanced Diet Formulation and Preparation of Feeds for Rats

Nutrient Requirements of Rats

Nutritional requirements can be worked out by using several criteria and requirement for a given nutrient may vary with criterion used. Growth, reproduction, nutrient storage, enzyme activity, gross and histological appearance of tissues and their content of nucleic acid and proteins are major criteria used to assess the nutritional adequacy of a diet.

Growth and Reproduction

Average daily intake of diet is approximately 15-20g in males and 10-15g in females. Pregnancy raises intake to about 20g and lactation to 30-35g.

Fats

Fat is considered as optional component of diet except as a source of essential fatty acids and its role in utilization of dietary fat soluble vitamins. Out of the 3 essential fatty acids, arachidonic acid has highest biological potency for growth. Arachidonic acid is found predominantly in animal tissues. Linoleic acid is widely distributed in plant oils and is most abundant dietary essential fatty acid. Arachidonic acid itself is not essential and it can be derived from linolenic acid.

Carbohydrates

Although it was thought that there was no specific requirements of carbohydrates it is clear that glucose or glycogenic precursor should be included in the diets of rat. Diarrhoea may occur if dietary sugar content is high.

Fibre

The usual practice is to include 5% of non-nutritive fibre in the diet. Excessive dilution of diets with fibre decrease nutrient and calorie intake.

Protein and amino acids

Growth

In establishing protein requirements for each of the physiological function, 3 factors have to be considered (i) energy content of diet (ii) amino acid composition of protein (iii) digestibility of protein.

Since the feed intake is related to energy content of diet, the protein requirement can bé expressed as protein to calorie ratio. About 22mg of apparently digested protein/Kcal of apparently digested energy produces most efficient nitrogen retention. If ingredients are selected to provide adequate amounts of essential amino acids 15-20% total protein in sufficient.

Maintenance

The dietary requirement is 10.7 mg protein per gross kcal, assuming a gross kcal requirement of 117/kg $W^{0.75}$ with 200mg N multiplied by 6.25.

Gestation and lactation

The net protein requirement for gestation and lactation as percent diet does not differ significantly from that of the growth of weanling rats. The requirement of protein for growth, gestation and lactation 12% of air dried diet as net protein. For maintenance of adult rats, it is 4.2% as net protein.

Energy

Most purified, low fat (5-10 percent) diets contain from 4.0 to 4.5 Mcal/kg of gross energy. DE of purified diets ranges from 90-95% of GE and the ME varies from 90 to 95% DE.

Maintenance

The maintenance energy requirement is the sum of needs for basal heat production, physical activity, alterations in ambient temperature, and those needs associated with the consumption and utilization of food energy. Animals fed at maintenance level should maintain energy equilibrium. The daily maintenance energy requirement in ME units for adult rats is 114 Kcal of ME/kg $W^{0.75}$.

Growth

Weanling and mature rats consume about 225 and 150 Kcal DE/day/kg $W^{0.75}$, respectively.

Gestation and Lactation

The energy requirement for gestation appears to be 10-30 percent greater than that of mature female rat fed *ad libitum*. Food intake of rats fed diets adequate in protein is increased by 20-30 percent during the first day of gestation and up to 140 percent by 16th to 18th day of gestation. Protein appears to be more critical than energy for satisfactory reproduction. Lactating rats consume from 2 to 4 times the energy of non lactating females.

The nutrient requirements of rats as suggested by NRC, (1978) are given below.

Nutrients Requirements of Rats

Nutrient	Concentration in a diet*	
	Growth, Gestation or Lactation	Maintenance
Protein %	12.00	4.20
Fat ** %	5.00	5.00
Digestible Energy Kcal/kg	3800	3800
Calcium %	0.50	
Phosphorus %	0.40	

* Adequate to support growth, gestation and lactation (based on 90% dry matter)
** Linoleic acid, 0.6 percent is required

BIS specifications for compounded feeds for laboratory mice and rats

S.No	Characteristic	Requirements
1.	Moisture (%) max	10
2.	Crude protein (min)	24
3.	Crude Fat (max)	5
4.	Crude fibre (max)	6
5.	Total ash (max)	9
6.	Acid insoluble ash (max)	1
7.	Calcium (min)	0.6
8.	Available phosphorus (min)	0.3

Suggested ration for rat

Ingredient	Quantity (g)	protein %	Energy (Kcal)
Wheat flour whole (12, 345)	300	36	1035
Maize flour (11, 342)	300	33	1026
Bengal gram flour (roasted) (22.5, 370)	200	45	740
Whole milk powder (26, 496)	100	26	496
Wheat bran (13,250)	85	11.05	212.50
Mineral mixture	10		
Salt	5		
Total	1000	151.05	3509.50

Vitamin Mix. – 10 g/kg feed.

The protein content of above diet is 15% and the energy content is 3509.5 Kcal (GE = 3334 Kcal ME).

Exercise

1. Formulate a ration for growing rats by using locally available feed ingredients.

15. Balanced Diet Formulation and Preparation of Feeds For Rabbits

The domestic rabbit (*Oryctolagus cuniculus*) is primarily herbivorous and will consume most types of grains, greens and hay. Diets provided, whether home grown or commercially prepared, consist almost entirely of ingredients from plant sources. Rabbits habitually practice coprophagy, sometimes referred to as pseudo rumination.

Fermentation in the large intestine and the practice of coprophagy probably provide the necessary amounts of most B vitamins, provide some bacterially synthesized protein and may permit further digestion of some nutrients by multiple passages through the digestive tract. The high digestibility of forage protein in rabbits is partially due to coprophagy. Mature rabbits vary in size from 1 to 6 kg. The nutrient requirements of rabbits published by NRC (1977) are presented in Table 1.

Table 1: Nutrient Requirements of Rabbits fed *Ad libitum* (percentage or amount per kg of Diet)

Nutrients	Growth	Maintenance	Gestation	Lactation
Digestible Energy (Kcal)	2500	2100	2500	2500
TDN (%)	65	55	58	70
Crude fibre (%)	10-12	14	10-12	10-12
Fat (%)	2	2	2	2
Crude protein (%)	16	12	15	17
Calcium (%)	0.4	--	0.45	0.75
Phosphorus (%)	0.22	--	0.37	0.50

Energy

Approximately 9.5 Kcal of digestible energy (DE) is required per gram body weight gain, regardless of energy content of the diet. A level of 2,500 Kcal of DE per kg of diet will satisfy the energy needs for rapid growth. For an adult rabbit of 3 kg the daily metabolizable energy requirement is 200 kcal, a quantity easily provided with diets of 2,100 to 2,200 Kcal of DE/kg.

Protein and Amino Acids

For rapid growth, rabbits are dependant upon adequate quantities of dietary essential amino acids. Bacterial protein synthesis in the caecum has been demonstrated, but this protein obtained by means of coprophagy, apparently does not make a large contribution to the essential amino acid needs of the young rabbit. Non protein nitrogen sources would not be useful to rabbits. Crude protein levels of 16, 12, 15, and 17 per cent are recommended for growth, maintenance, pregnancy and lactation, respectively.

Example of adequate diet for commercial production

Formulation of ration for growing rabbits (0.5 to 4 kg) to meet the requirements for their growth.

Requirement: Protein – 16% Energy – 65%

Ingredient	Quantity (%)	Protein (%)	Energy (%)
Ground Lucerne hay (18,55)	20	3.6	11
Ground Maize (9,80)	35	3.15	28
Wheat bran (13,57)	30	3.9	17.1
Groundnut cake (45,80)	12	5.4	9.6
Mineral Mixture	2	--	--
Salt	1	--	--
Total	100	16.05	65.70

Vitamin mixture @ 20g/100kg

Exercise

1. Formulate a ration for maintenance of an adult rabbit using locally available feedstuffs.
2. Formulate a commercial ration for lactating does using local feedstuffs.
3. Formulate a diet that can be used to maintain pregnant does by using locally available feedstuffs.
4. Visit the local rabbit farm, observe and note the type of feeding schedule of rabbits and composition of various diets that are fed to bunnies, bucks and does (lactating and pregnant).

16. Balanced Diet Formulation and Preparation of Diets for Guinea Pigs

The domestic guinea pig (*Cavia Procellus*) has been bred in captivity for at least 400 years and originated in Peru, Agrentina or Brazil. It is a monogastric herbivore, and the gastric mucosa does not have a keratinized portion. Another notable feature is the presence of large semi-circular caecum with numerous lateral pouches. The organ resembles that of the rabbit and possibly has similar digestive functions e.g. synthesis of B vitamins by microorganism and recycling of intestinal contents by coprophagy.

New born animals can consume semi-solid and solid food immediately. In its native habitat, the guinea pig diet consists of green vegetation and fruits; whereas in the laboratory the diets are generally of much higher caloric density and lower fiber content. Furthermore, the guinea pig eats many small meals throughout the day, may be selective in its choice of diets, and may resist abrupt changes in formula or form of diet. Palatability can be improved by pelleting.

The guinea pig is best known from a nutritional stand point because of its requirement for dietary vitamin C. This feature has made them particularly useful in studies of collagen synthesis, wound healing and bone growth. The growth of guinea pig has a relatively high requirement for arginine and folic acid.

Energy

Digestible energy of diets of guinea pigs should be 3.0 to 3.25 kcal/g of dry matter.

Protein

Plant proteins contain generous amounts of arginine, and the guinea pig, an herbivore, grows well on diets that contain 18 per cent protein from plant sources.

Vitamin C

Daily intakes of 6 to 10 mg vitamin C will provide adequate amount for growth and reproduction: normal intake of a diet with 200mg/kg will fulfill this need.

The recommended nutrient allowances (NRC, 1978) for growing guinea pigs are given in Table 1.

Table 1: Recommended Nutrient allowances for growing guinea pigs

Nutrient	Amount in diet
Protein	18%
Unsaturated	% 1% corn oil satisfactory
Digestible Energy	3 kcal/g
Fiber	10%
Calcium	0.8 to 1.0%
Phosphorus	0.4 to 0.7%
Vitamin C	200 mg/kg

Table 2 : Composition of a natural ingredient diet used for Guinea pigs (NRC, 1978)

Nutrient	Level (%)
Alfalfa meal	38.15
Ground wheat	28.90
Ground oats	17.75
Soybean meal	13.25
Ground lime stone	1.10
Iodized salt	0.50
Dicalcium phosphate	0.25
Mineral mixture	0.05
Vitamin mixture	0.05
Total	100.00

For more details on purified diets used for guinea pigs, the student is advised to refer 'Nutrient Requirements of Laboratory Animals, (1978) published by National Academy of Sciences, Washington, D.C.

Exercise

1. Formulate a ration of growing guinea pigs using locally available feedstuffs.
2. Visit the local 'Laboratory Animal House' and note down your observations on the feeding of various lab animals.

17. Hygienic Preparation of Balanced Diets: Principles and Guidelines

Food is classified into 3 groups on the basis of spoilage. Spoilage is a condition produced by excessive growth of micro organisms leading to decay or decomposition. The non-perishable foods like flour, by-products and dry grains ordinarily do not spoil unless handled carelessly. The perishable foods like meat, fish, poultry spoil readily unless preservative methods are used. The semi perishable foods like fruits and vegetables having thick skin can be protected from spoilage if handled properly. Spoilage is caused by moulds, yeast, bacteria and insects.

Moulds

Foods when affected by moulds are covered with white or graying organic matter. They are members of plant kingdom. They are devoid of chlorophyll and consist of mycelium and spores. The spores are resistant to drying and light and multiply on exposed food. They thrive on closed damp situation. Acid medium is favourable. The moulds consume nutrients present and lower the food value. Eg; Aspergillus, Penicillium, Rhizopus etc.

Yeast

It is non-motile, round, oval unicellular organisms. They multiply by budding once grow under moderate temperature in solutions containing sugar and plenty of water. Yeasts are very useful in making bread, beer, wine, vinegar and many other fermentation products. They are undesirable when they grow on fruits, fruit juices, squashes, sharbat, honey etc., like moulds they also destroy the appearance and wholesomeness of fruits and fruit products. Bubbling in fermentation products is the indication of presence of yeast. Boiling destroys yeast, e.g: Candida, Saccharomyces etc.

Bacteria

Most common cause of spoilage: Bacteria thrive on nitrogen rich foods while moulds and yeast prefer sugar and acid food materials. Diseases like typhoid, tuberculosis, diphtheria are caused by bacteria. Enormous economic losses are encountered since food is made unfit for consumption when spoiled by bacteria. Anaerobic bacteria cannot grow in the presence of oxygen. The growth of bacteria is controlled by a number of external factors. They must have proper kind of food material, heat, pH reaction, moisture and oxygen. If any of these factors are limited, growth is retarded to some extent. Low temperatures do not kill bacteria but check or stop their growth and as soon as the temperature is raised above the minimum point, they resume growth. Some bacteria are heat resistant. Moist heat is fatal to their growth. They can be killed by heating at a temperature of 121°C for 30-40 min. *Clostridium botulinum* is responsible for spoilage of canned foods and is highly poisonous.

Enzymes

Enzymes produce normal ripeness of fruits and vegetables but if their activity is not checked, they cause staleness and decay. The turning of the cut or the wounded surface of apples to brownish colour on exposure to air is also due to an enzymatic action. Enzymes can be easily inactivated by heat.

Insects

Worms, bugs and fruit flies may damage food and render it unfit for human consumption. The injuries caused by these insects serve as pathways by which micro-organisms reach the inner tissues. Dehydrated fruits and vegetables are more commonly infested by insects etc. during storage.

Principles and Methods of Preservation

Food spoilage may be caused due to chemical and biological factors. The most significant changes in food are due to biological factors. Inherent enzymes that are normally present in food under certain conditions cause spoilage. But the presence and growth of bacteria, mould and yeasts in food are more serious in spoilage than the action of inherent enzymes. Micro-organisms are present in the air, water on most foods and our hands, in fact they are ubiquitous. In accomplishing the preservation of foods using various methods the following principles are involved.

1. Prevention or delay of microbial decomposition

Many methods of food preservation depend not on the destruction or removal of micro-organisms rather on inhibition of growth and hindrance to growth once it has begun. This can be achieved by keeping out the micro-organisms (asepsis), removal of micro-organisms (filtration through bacteriological filters), hindering the growth of micro-organisms (low temperature, drying, anaerobic conditions, or use of chemicals) and by killing the microorganisms (heat and radiations).

2. Prevention or delay of self decomposition of foods

This can be achieved by destruction or inactivation of food enzymes by heating, which not only destroys the enzymes but eliminates micro-organisms. Self decomposition can also be prevented by the action of purely chemical reaction e.g. prevention of oxidation by means of an anti-oxidant.

3. Prevention of damage by insects etc.

Food if not packed and stored properly is attacked by insects and worms etc. which damage the products and make unfit for human consumption. This can be prevented by the use of chemicals and fungicides.

Methods of preservation

The chief methods of food preservation are:

i. Asepsis
ii. Filtration and pressure
iii. Fermentation
iv. Heat processing
v. Refrigeration
vi. Drying
vii. Preservative
viii. Radiation
ix. Antibiotics

i. Asepsis

Preservation is mainly based on the keeping out of micro-organisms. The healthy inner tissues of fruits and vegetables are free from micro-organisms, and if there is a protective covering about the food, microbial decomposition is delayed or prevented. Packaging of food is widely used to attain asepsis. The covering prevents primary contamination during handling, to the hermetically sealed containers of canned foods, which if tight, protects the processed contents from contamination by micro-organisms.

ii. Filtration and pressure

Removal of micro-organisms is not very effective but under special conditions may be useful. Removal may be accomplished by filtration, centrifugation, washing or trimming. The liquid is filtered through a previously sterilized "bacteria proof" filters made of asbestos pad, sintered glass, diatomaceous earth, unglazed porcelain, similar material, and the liquid forced through by positive or negative pressure. This method has been used successfully with fruit juices, beer, soft drinks, wine and water. Similarly washing trimming of fruit and vegetables prior to processing reduces the land of micro-organisms.

iii. Fermentation

This is one of the oldest methods of preservation. Decomposition of carbohydrates by micro-organisms or enzymes is called fermentation. By this method, the foods are preserved by organic acid formed by microbial action. Alcoholic, acetic and lactic types are three important kinds of fermentations in the food industry. The keeping quality of alcoholic beverages, vinegars and fermented pickles depends upon the presence of alcohol, acetic acid and lactic acid, respectively. Care should be taken to seal the fermented products from the air to avoid further unwanted or secondary fermentation, Wines, beers; fruits vinegars, fermented drinks, fermented pickles etc. are prepared by these processes. 14 per cent alcohol acts as a preservative in wines because yeasts etc. cannot function in that concentration. 1.8 per cent lactic acid and 2.5 per cent salt act as a good preservative in fermented pickles.

iv. Heat Processing

The term processing or sterilization by heat means the complete destruction by heat of all forms of life in the product sterilized and is applied to solid canned packs of fruits and vegetables. In commercial practice not all cans of food are sterile. But they usually do not spoil because the conditions in the can are not favourable for the growth of the organisms concerned. The acidity may be high or oxygen may be lacking. Therefore, the term "processing" is preferable to the term "sterilization" in case of canned foods.

Heat treatments which are employed in processing foods are:

a. *Pasteurization*: This term has been coined after the name of Louis Pasteur, a great French chemist and Bacteriologist. It is a heat process that not only kills many organisms but also greatly weakens those not killed. This also inactivates the enzymes which may otherwise cause spoilage. Pasteurization refers to the heat treatment applied to acid and some non-acid liquid products like fruit juices, vinegar, wines, beers, milk, sauces, carbonated soft drinks etc and usually involve the application of temperature below 100°C. The heating may be done by means of steam, hot water and dry heat. Time and temperature used in the pasteurization process depend upon the method employed and the product treated. The high temperature-short time (HTST) method employs a comparatively high temperature for a short time. The low temperature long time (LTLT) or holding methods use a lower temperature for long time.

b. *Heat at about 100°C* : Most fruits and some of the vegetables like tomatoes and rhubarb etc. are acidic and therefore, their canned products are easily preserved by processing at 100°C. A temperature of approximately 100°C is obtained by boiling a liquid food, immersion of the container of food in boiling water, or by exposing to steam. Blanching of fresh vegetables before freezing or drying also involves heating briefly at about 100°C.

v. Refrigeration

Low temperatures are used to retard chemical reactions, action of food enzymes and to slow down or stop growth and activity of microorganisms in food. The lower the temperature, the slower will be chemical reactions, enzymatic actions and microbial growth of any micro-organisms. Temperatures that are just above freezing maintain foods near their original conditions are more costly and may result in undesirable changes in some foods. Some bacteria mould and yeast can grow slowly at temperatures several degrees below the freezing point of water. Therefore, even a temperature of 0° C or slightly below will not prevent microbial spoilage of most raw foods indefinitely. Low temperature storage can be done in three ways (a) common or cellar storage (b) chilling and (c) cold storage or freezing.

vi. Drying

Drying usually is accomplished by removal of water, but any method that reduces the amount of available moisture in a food is a form of drying. Moisture may be removed from foods by any number of methods. In sun dried food moisture is removed by exposure to the sun's rays without any artificially produced heat and without controlled temperatures, relative humidities, or air velocities for dehydration food is dried by artificially produced heat under controlled conditions of temperature, relative humidity and air flow. If the moisture has been removed from a liquid food it is called condensed food. Preservation by drying depends upon reducing the moisture content to the point at which the concentration of dissolved solid in the product is about 70 per cent or above and this prevents not only the growth of spoilage organisms but also inhibits the action of enzymes.

vii. Preservatives

"Preservatives" means a substance which when added to food is capable of inhibiting, retarding or arresting the process of fermentation, acidification or other decomposition. It may be a chemical compound or mixture of compounds applied for specific purpose of preventing spoilage due to the growth of bacteria, yeast or mould etc.

Classification of Preservatives

Sugar, common salt, vinegar or acetic acid etc, are termed as household preservatives. Their functions in food preservation are as under:

a. *Sugar*: Sugar used at a concentration of 65 per cent or above permanently preserves most foods as syrups, jellies and jams. It acts by osmosis.

b. *Salt*: Salt acts by osmosis. In concentration of 10-15 per cent it is sufficient to preserve most products. It inhibits enzymatic browning, discolouration and acts as an antioxidants too. It is also used in curing of fish and pork. Salt in the form of brine is used for canning and pickling of vegetables.

c. *Vinegar or acetic acid*: Vinegar or acetic acid checks aerobic and anaerobic fermentation. It is a poison for microbial spoilage and possesses germicidal and antiseptic properties. About 2 per cent of acetic acid prevents the spoilage of most products. It is used in the preservation of pickles, sauces, chutneys, relishes, meat, fish etc. It is also a useful tool to modify taste and flavour.

d. *Benzoic acid and its salts*: Benzoic acid is sparingly soluble in water. Hence sodium benzoate as its salt is used. It is 180 times more soluble than the acid. The salt is tasteless, odourless and is an organic antagonist. It is the benzoic acid molecules which is germicidal. Its conservation action increases in the presence of carbon dioxide and acid media. It is more effective against yeasts than against moulds. 0.06-0.10 per cent of sodium benzoate preserves most fruit products (pH 3.5-4.0). The action of the benzoate is diminished markedly at pH 5.0 and therefore for preservation in soft drinks 350 ppm of sulphur dioxide and 600 ppm of benzoic acid are permitted. In juices and soft drinks 0.1 per cent of benzoate may cause a disagreeable burning taste. In the long run, the benzoate may darken the products. It is therefore, mostly used in coloured products of tomato, phalsa, plum, watermelon, jaman, pomegranate, coloured grapes etc., where use of sulphurous acid causes fading of the organic colour (anthrocyanin pigment).

e. *Sulphurous acid and its salts*: It is mostly administered in the form of its salt i.e. potassium or sodium metabisuphite. It is crystalline salt and gives characteristic sulphur dioxide smell. Sulphur dioxide acts mostly as a preservative, retards oxidation, prevents discolouration or loss of flavour and of vitamin C. It provides better preserving action against bacterial fermentation and moulds. Being highly soluble it ensures better mixture and is an effective preservative. It prevents darkening of pared fruits and vegetables due to enzymatic action. It cannot be used in products stored in tin cans because it causes pin holes in metals and forms ugly encrustation in the form of tin sulphide. In dehydration of fruits and vegetables burning of sulphur at levels 1000-3000 ppm preserves colour, vitamin C, repels insects and destroys organisms in fruits and vegetables. It bleaches colour of anthrocyanin pigment and, therefore, its use may be restricted to the products of fruits like mango, litchi, lime, oranges, lemon, bael, guava etc., at 350 ppm. This works out to be 0.6gm of bisulphate per kg of the finished product. In raw juices to be stored for making squashes in the off season 700 ppm of sulphur dioxide may be used.

f. *Sorbic acid*: It is an organic acid having antimicrobial properties. In the pure form it is white crystalline powder, slightly acidic with agreeable taste. It is a harmless food preservative and prevents moulds in preserves. The effectiveness of sorbic acid increases in acid media (less than pH 6.0) and at this condition inhibits moulds and yeasts as well as some bacteria. Its level of use in jams, fruit syrups and pastes is 0.025, 0.02 and 0.10-0.02 per cent, respectively.

Sodium and potassium salts of sorbic acid are used as fungi static agents for foods especially on the surface and in the wrapping materials. These are also used for the preservation of cheese, sweet pickles etc and for the control of lactic acid fermentation.

viii. Radiation

Although, preservation of selected food products by radiation has been successful but there are still doubts about the safety of these products for human consumption, radioactive substances give off rays which are invisible to the naked eye but can be detected by a photographic plate. Three kinds of rays emitted from radioactive substances are alpha, beta particles and gamma rays. These are also called ionizing radiations. Therefore, sterilization by means of these radiations sometimes is termed "Cold sterilization". The dosages of radiation necessary for complete sterilization of a product usually do not raise the temperature to more than 28°C. Thus packaging should be much simpler and more economical. Radiation may also produce physical or chemical changes in the food, causing alteration of flavour, colour and sometimes texture of the food.

ix. Antibiotics

An antibiotic is a chemical substance produced by micro-organisms which have the capacity of inhibiting growth or destroying other kinds of micro-organisms. Their use and application in medicine is well known in controlling certain disease producing organisms in the body. Some antibodies are also in use to preserve the fruits, vegetables and their products. Subtelin, an antibiotic obtained from certain strains of *Bacillus subtilis* is used to facilitate preservation of asparagus, corn and peas. It is most effective against gram positive bacteria and spore forming organisms. Canned peas and tomatoes containing 10 and 20 ppm of subtelin were free from any growth of micro-organisms. Subtelin and nisin effectively reduce the thermal process requirements necessary to control the spoilage of several food products.

Nisin is another antibiotic produced by *Streptococcus lactie* organisms commonly found in milk, curd, cheese and other fermented milk products. Nisin is non-toxic and has no adverse effect on the sensory qualities of food. Nisin is widely used in food industry and in canned food. It is especially useful in acid foods in which it is more stable. It is more commonly used in canning of mushrooms, tomatoes and milk products. Nisin suppresses the growth of spoilage organisms mainly the gas producing, spore forming bacteria and toxin producing *Clostridium botulinum*.

18. Storage of Feeds

Most of the ingredients that are used for compounding a diet are liable to undergo deterioration on storage, especially in adverse storage conditions. All diets therefore should be used within a short time after preparation. Deterioration is caused by infestation with a variety of arthropod pests and mites. Some of these pests may even act as carriers of eggs of some parasites e.g. weevils are carriers of ova of tape worm. Fungi and moulds will multiply in damp conditions, especially in worm temperature. Wild rats and mice will be attracted to food stores and contaminate the feed and cause spready of several infections.

It is therefore necessary to store feeds in a cool, dry, vermin- proof rooms. Pelleted feeds should be packed in multiplayer bags containing one inner polythene layer which has been heat sealed and is air tight. All bags to be stored should be unbroken. Each bag should bear the date of manufacture and batch number. The feed bags for storage should be arranged in such a way that feed manufactured at an earlier date are consumed first. Dietary ingredients should be subjected to physical examination upon its arrival in the store, and if found inferior in quality, it should be rejected out right. Contaminated ingredients can be easily identified by the presence of clumps, faecal droppings or arthropod pests such as flour moth, beetles, weevils and mites.

In case of such contamination, it is to reject such ingredients. As a matter of convenience, ingredients required for a period up to 3 months can be procured at one time, stored in ideal conditions, and used for compounding diet. The feeds must be screened and kept free from salmonella infection.

❑❑❑

INDEX

D

E

F

G

H

R

S

T

U

V

W

□□□

About the Authors

Dr. S.P.Tiwari is Professor and Head, Department of Animal Nutrition at College of Veterinary Science and Animal Husbandry, Anjora, Durg, under the aegis of Indira Gandhi Krishi Viswavidyalaya, Raipur, Chhattisgarh. Dr. Tiwari did his BVSc & AH from College of Veterinary Science and Animal Husbandry, Mhow, in 1981 and received Maj. Juneja Trophy for getting highest marks in the faculty as "Best out going student". Dr. Tiwari did his MVSc in Animal Nutrition from Haryana Agriculture University, Hissar in 1983. He has completed his PhD in Animal Nutrition from Indian Veterinary Research Institute, Izatnagar in 1988. He started his career as Assistant Professor in Himachal Pradesh Agriculture University, Palampur from 1988 to 1991 and subsequently Associate Professor at Rajasthan Agriculture University, Bikaner from 1991 to 1993. Thereafter, he joined in the same capacity at Indira Gandhi Agriculture University, Raipur from 1993 to 1999. Dr. Tiwari joined the same University as Professor of Animal Nutrition in 1999 and Head of the Department in 2004 and continuing in the same capacity at the College of Veterinary Science & AH, Anjora, Durg. Dr. Tiwari is a distinguished Animal Nutritionist and contributed significantly in the area of strategic supplementation, integrated farming system, Medicated urea molasses mineral blocks and clinical nutrition. He received ND Kehar Award for the best PhD thesis. He has guided 20 MVSc and 3 PhD students. Dr. Tiwari was PI and Co-PI of many ICAR adhoc research projects besides NARP, NATP and State and other University level projects. He has published about 110 research papers and three book chapters on various aspects of animal nutrition. He was awarded with Merit scholarship during undergraduate and Junior and Senior Research Fellowships during masters and doctorate degree programme by Indian Council of Agriculture Research, New Delhi. He is life member of ANA, ANSI, ISNA, IPSA, IAVNAW and many other scientific societies. He is Fellow of Animal Nutrition Association and Fellow of Indian Poultry Science Association. He is member of Academic council and Administrative council of the university.

Dr. S.Rajagopal did his BVSc degree in 1963 from S.V. University Tirupati. He served in the various capacities the Animal Husbandry Department, Govt.of Andhra Pradesh from 1963 to 1976. He obtained his Masters degree in 1971 and Doctorate degree in 1976 in Animal Nutrition from Agra University, Agra. Thereafter he joined ASRB as Scientist in the year 1976. During his tenure as Scientist he was closely associated with other scientists at IVRI, Izatnagar in establishing the animal calorimeter and respiration chamber. He served as Associate Professor in G.A.U. S.K. Nagar, Gujarat in the year 1982 and established the Animal Nutrition Labs. In the year 1996 he joined as Professor & Head, Animal Nutrition, in the College of Veterinary Science & A.H., Anjora, Durg under I.G.K.V. Raipur. During his tenure he was the Principal Investigator for N.A.T.P. Project funded by ICAR, New Delhi. After attaining the superannuation retired as Professor & Head in the year 2004.

Dr. (Mrs.) Usha Rani Mehra did her graduation and post-graduation from Bareilly College, Bareilly in 1961 and 1664, respectively and doctorate in animal nutrition from Indian Veterinary Research Institute, Izatnagar in 1968. She started her research career in 1968 at Indian Veterinary Research Institute, Izatnagar and served the institute for four decades in various capacities till her superannuation as Head Professor cum Director, Centre for Advanced Studies in Animal Nutrition. Dr. U.R. Mehra is a distinguished Animal Nutritionist and contributed more specifically in the area of urea molasses mineral block, liquid diet, bypass protein sources, microbial growth and VFA production. She was bestowed with several prestigious awards like Hari Om Ashram Trust Award, Punjab Rao Desmukh Best Woman Agricultural Scientist Award, Rafi Ahmed Kidwai Award and Women Empowerment and Livelihood Award. She has published 180 research papers and many book chapters on various aspects of animal nutrition. She has guided several MVSc and PhD students. She is a Fellow of the National Academy of Agricultural Science (India) and a life member of many scientific associations and served as member of several important Academic Bodies, Research Panels, Research Advisory Committees and Board of Management. She provided dynamic leadership to ANA as its President from 2002 to 2005. She also served as Scientist Emeritus for two years in the Center of Advanced Studies in Animal Nutrition, after her superannuation at IVRI, Izatnagar.